D0603585

CRIME SCENE
INVESTIGATION

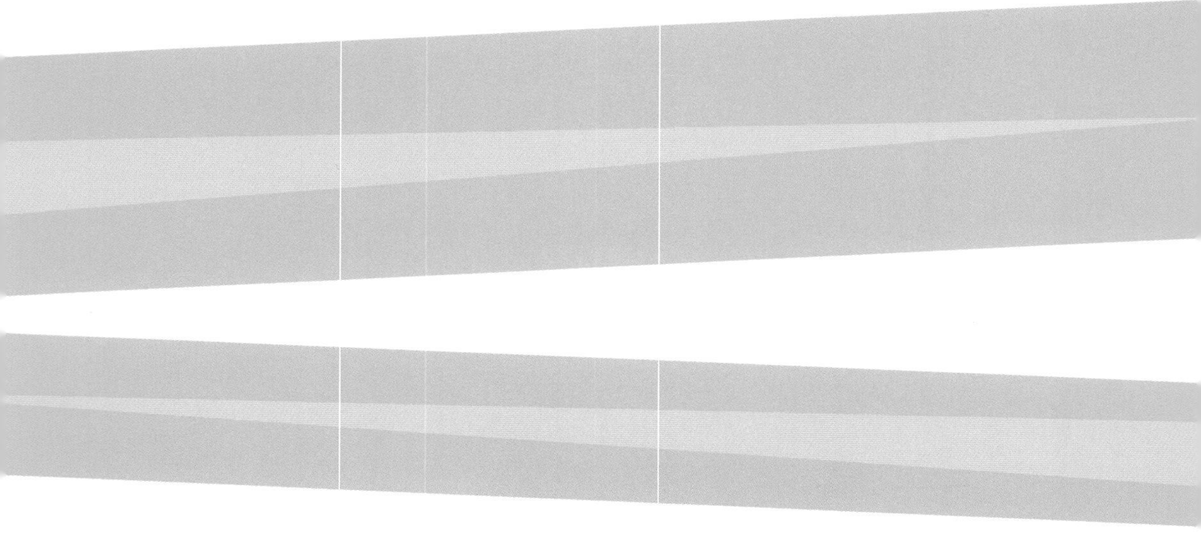

CRIME SCENE

Introduction by
Cyril H. Wecht, MD, JD

INVESTIGATION

Crack the case with real-life experts

Joseph T. Dominick, RN, LFD
Steven A. Koehler, MPH, PhD
Shaun Ladham, MD
Thomas Meyers, BS, MS
Timothy Uhrich, JD
Cyril H. Wecht, MD, JD
Michael Welner, MD

The Reader's Digest Association, Inc.
Pleasantville, New York/Montreal

A READER'S DIGEST BOOK

This edition published by The Reader's Digest Association Inc., by arrangement with Elwin Street Limited

Conceived and produced by
Elwin Street Limited
79 St John Street
London EC1M 4NR
United Kingdom
www.elwinstreet.com

FOR ELWIN STREET LIMITED
Managing Editor: Jane Morrow
Editor/Researcher: Ian Penberthy
Designer: Sharanjit Dhol
Illustrator: Richard Burgess

FOR READER'S DIGEST
U.S. Project Editors: Kim Casey, Nancy Shuker
Canadian Project Editor: Pamela Johnson
Project Designer: George McKeon
Executive Editor, Trade Publishing: Dolores York
Associate Publisher, Trade Publishing: Christopher T. Reggio
Vice President & Publisher, Trade Publishing: Harold Clarke

Library of Congress Cataloging in Publication Data:

Crime scene investigation : crack the case with real-life experts / general editor Cyril H. Wecht ; contributors Steven A. Koehler ... [et al.].
p. cm.
Includes index.
ISBN 0-7621-0540-2
1. Crime scene searches. 2. Criminal investigation. 3. Evidence, Criminal. I. Wecht, Cyril H., 1931– II. Koehler, Steven A.

HV8073.C6927 2004
363.25'2–dc22 2004050830

Address any comments about *Crime Scene Investigation* to:
The Reader's Digest Association, Inc.
Adult Trade Publishing
Reader's Digest Road
Pleasantville, NY 10570-7000

For more Reader's Digest products and information, visit our website:
www.rd.com (in the United States)
www.readersdigest.ca (in Canada)

Printed in Singapore

1 3 5 7 9 10 8 6 4 2

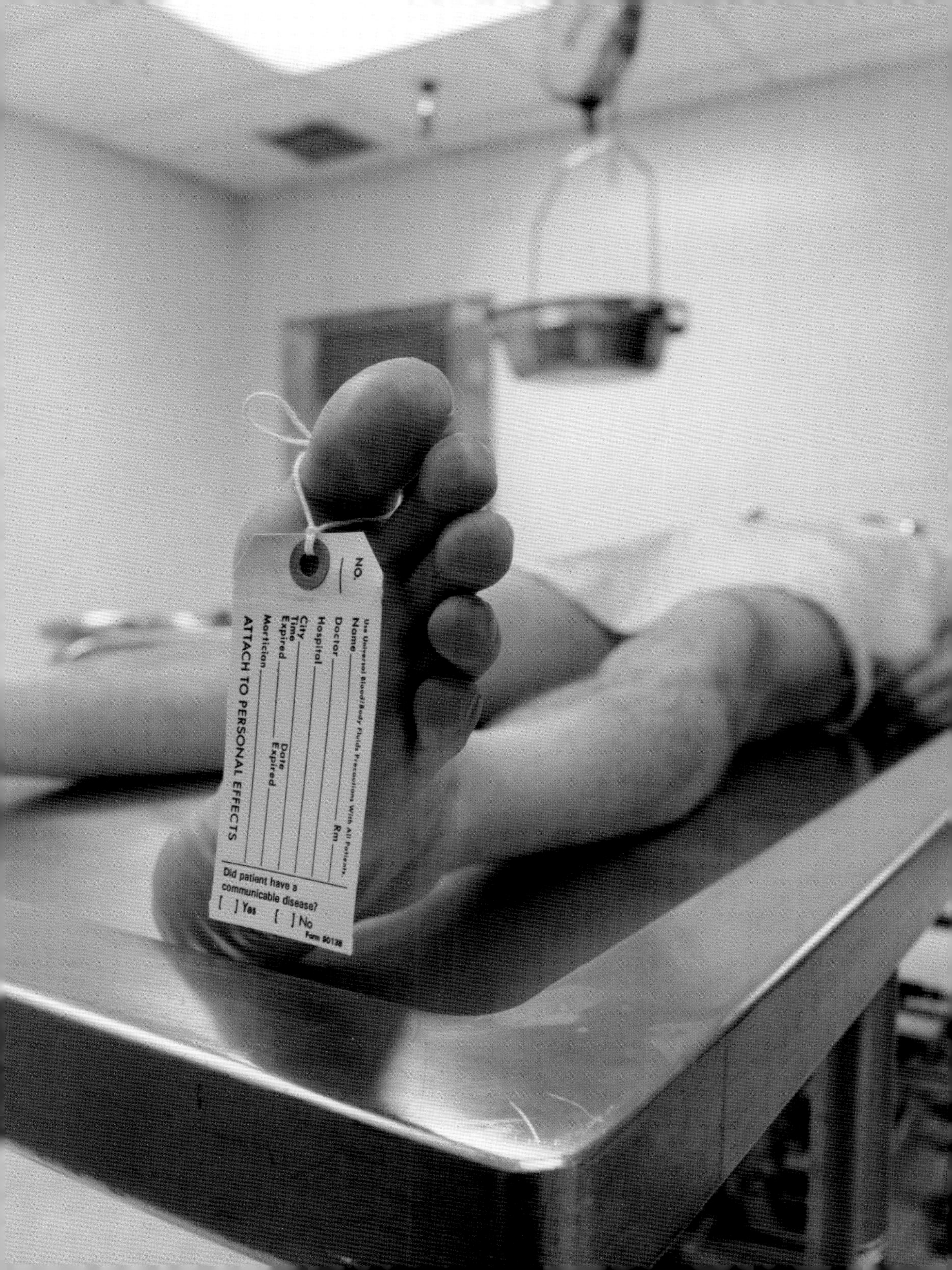
NO.
Use Universal Blood/Body Fluids Precautions With All Patients.
Name
Doctor
Rm
Hospital
City
Time Expired
Date Expired
Mortician
ATTACH TO PERSONAL EFFECTS
Did patient have a communicable disease?
[ ] Yes [ ] No
Form 90138

Contents

Introduction

It is difficult to think of any professional field in modern times that has had the intense and widespread impact upon society that we have seen in forensic science.

The current intense interest in forensic science has been fueled by the success of recent TV shows such as *Crossing Jordan*, and *Forensic Files*. Sherlock Holmes, Hercule Poirot, and Charlie Chan have given way to *Quincy* and *CSI*. This book takes you beyond fiction and into the real-life world of the homicide investigator questioning witnesses: the criminalist collecting hair, fibers, blood, and shoe prints at the death scene; the forensic pathologist examining the body in the morgue; the toxicologist analyzing body fluids; and various other members of the forensic scientific investigative team as they assemble all the evidence to identify the assailant, convict the guilty party, or exonerate the innocent. While the science is complex and multifaceted, *Crime Scene Investigation* will allow you—with or without a scientific background—to explore and appreciate this fascinating subject.

The fictional dramatization of highly skilled forensic scientists solving complex murder cases and other major crimes has been a important factor in fueling the popularity this field currently enjoys throughout much of the world. Another important factor responsible for this phenomenon is the occurrence of the highly controversial, notorious, and well-publicized real-life cases such as John F. Kennedy's assassination, the death of Marilyn Monroe, the O. J. Simpson case, the unresolved mysterious death of JonBenet Ramsey,

PREVIOUS PAGE Homicide investigators attend to a death scene. They will determine if it was an accident or murder.

ABOVE John and Patsy Ramsey respond to the media through their lawyer, Lin Wood, after being questioned by Boulder, Colorado, Police concerning the death of their daughter, JonBenet, in 1997. They denied having any part in the murder of their daughter. The case remains unsolved.

and many others that have made millions of people aware of the significant and essential role that forensic scientists play in criminal investigations. Actual trials are often reported in great detail by broadcast and print news media. Indeed, some cases are covered live, and many acquire an audience that rivals the following of the most popular TV soap operas.

Medicolegal Investigation and the Centuries-old Office of the Coroner

Although most people were not aware of the practical investigative significance of forensic science as recently as a generation ago, many references to this kind of work can be found in the chronicles of ancient societies.

To one degree or another, all civilizations have recognized the need for medicolegal investigation in their civil and criminal justice systems. The Code of Hammurabi, written in 2200 B.C., dealt in part with what is now called medical malpractice. The ancient Egyptians developed a system to determine the cause of death and whether it was natural. The Chinese compiled a volume titled *Hsi Yuan Lu* (*The Washing Away of the Wrong*), describing different procedures for investigating suspicious deaths.

In the Middle Ages, medicolegal investigation developed within two major systems. In continental Europe, the discipline always remained free from political influence;

objectivity and true expertise were maximized by the resultant autonomy. By the eighteenth and nineteenth centuries, many European universities had developed curricula in legal medicine.

In sharp contrast, the English system of medicolegal investigation was always an integral part of the political system. The office was established in 1194. Although not one of his responsibilities initially, the investigation of death soon became a function of the coroner. Subsequently, the duty was assigned to justices of the peace, but it was reacquired by the coroners in the late nineteenth century. At that time, the jurisdiction, which continues to this day, was first defined: the coroner was to investigate sudden, violent, or unnatural deaths and all deaths of prisoners.

In the United States there are two systems: a "coroner" system and a "medical examiner" system, and depending on the law in each state, one of these systems will be in operation. Both systems are derived largely from the British system. In the coroner system, the coroner is elected to the office, while the medical examiner is appointed (throughout this book, the term "coroner" appears interchangeably with "medical examiner").

In Canada, each province has its own legislation and courts and its own system of death investigation—some with a coroner system and others with a medical examiner system. A common feature in Canada is a hierarchical system, when a chief coroner heads the service with different levels of coroner beneath.

In Australia each state and territory has its own legislation, which is based in principle on the British system. The coroner has jurisdiction to hold an inquest when a person has died a violent or unnatural death; when the cause is unknown; when the circumstances are suspicious or unusual; and when the death wasn't attended by a medical practitioner.

The origin of the definitive link between law and medicine may be dated from about A.D. 530 and the publication of the Code of Justinian, which said that medical knowledge should be used in the adjudication of certain legal cases.

The Role of the Forensic Pathologist

The investigation of violent, sudden, suspicious, unexpected, unexplained, and medically unattended deaths to determine the cause, mechanism, and manner of death is the primary responsibility of the forensic pathologist. Quite often, it is also

The six Ws of forensic pathology

WHO	Sex, race, age, and particular characteristics of the victim.
WHEN	Timing of death and injuries.
WHERE	Scene and circumstances of death.
WHAT	Type, distribution, pattern, path, and direction of injuries.
WHICH	Significance of injuries—major vs. minor, true vs. artifactual or postmortem.
WHY	How injuries were produced—mechanism and manner of death.

necessary to ascertain the time and place of death; the relationship between natural disease and death; and, when two or more victims are found at the same location, the sequence of their deaths. In the United States, because of the burgeoning number of wrongful-death lawsuits, it is important to determine whether the victim experienced conscious pain and suffering before death. All of these findings require expertise in forensic pathology.

The consequences of such analyses and the conclusions and opinions expressed by experts are extremely important, not only for the individuals involved but also for the health, welfare, and safety of society in general.

Of course, the forensic pathologist cannot function effectively in a scientific vacuum. To make the kinds of thorough, complete, and accurate determinations that are essential to the success of an official medicolegal investigation, there must be timely and appropriate input from forensic toxicologists, chemists, serologists, and immunologists; forensic anthropologists, entomologists, and odontologists; and all the various subdivisions of criminalistics, including firearms and ballistics; hair, fibers and trace evidence; and fingerprint, footprint, and DNA analyses. In some controversial cases involving craniocerebral injuries, brain

A body awaits an autopsy in a morgue.

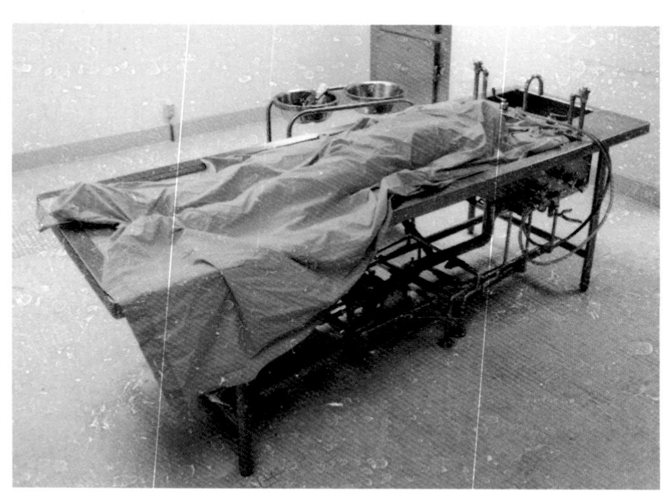

disease, and complex central nervous system disorders, forensic neuropathologists will also be consulted.

At the crime scene and afterward, specially trained homicide detectives are vital to ensure proper investigation. Skilled forensic photographers are also necessary to capture all the critical details of the victim's death on film and to supplement the gathered evidence by taking photos throughout the postmortem examination.

The Value of Expert Witnesses

The forensic pathologist is the medical expert who spends the greatest amount of time preparing for and giving testimony in court. Whether the death is the result of disease or violence, he or she is expected to determine the reason and testify to those findings in a court of law.

Deaths caused by poisoning, drowning, or any kind of trauma (recent or remote) are routinely studied by the forensic pathologist. In addition, battered-child syndrome has been catapulted into prominence in North America over the last decade, and the number of court cases dealing with this complex problem is growing steadily. The forensic pathologist

Investigators remove the body of Nicole Simpson on June 13, 1994. When the verdict was read during O. J. Simpson's trial in 1995, 142 million people listened in or watched it on television.

may also give evidence in cases dealing with rape, although other clinicians, particularly emergency-room specialists and gynecologists, are more likely to be involved. The identification of blood and other biological stains will result in the forensic pathologist, as well as the serologist, being called to court from time to time.

The role of the clinician in forensic medicine has greatly increased in importance over the past two to three decades. Although clinical forensic experts can provide a wide range of services regarding litigation, by far the most common function is serving as an expert witness to give testimony in a court of law. Other important functions include assisting attorneys in their investigations, helping to prepare the case, and interpreting the associated medical facts. Those areas of litigation that rely heavily on the experience and expertise of the forensic scientist have seen steady growth over the last few decades. There has been a recent marked increase in all Western countries in the number of personal injury actions, including, but not limited to, the rising number of medical malpractice claims filed. There has also been an increase in the number of product liability cases.

Even though the assassination of American President John F. Kennedy took place in 1963, forensic scientists are still able to return to the evidence collected to draw further conclusions and debunk conspiracy claims.

Since medical testimony or medical reports are important in the majority of court cases—over 80 percent of all cases litigated in the United States—it is likely that a significant number of physicians, regardless of their specialty, will be called upon to appear in court as expert witnesses.

There are many examples of civil and criminal lawsuits in which the failure of the trial judge and/or the attorneys to appreciate the significance of crucial forensic scientific evidence resulted in grave injustice or unresolved controversy. The value of consulting appropriate experts can only be fully comprehended if both judges and lawyers seek understanding of what each field of forensic science can offer.

Recent Technological Developments

Undoubtedly, the development of DNA testing would have to be considered the most significant and dramatic advance in forensic science. Its use in many kinds of homicide, rape, and sexual assault, and other criminal cases has added an entirely new dimension to medicolegal investigation.

Computer simulation of events is a relatively recent process that is being used in both criminal and civil cases. Another promising technique is image processing with computer analysis, which offers the potential of extracting information from photographs that cannot be gleaned by any other method.

The current terrorist threats from biological, chemical, and other weapons of mass destruction (WMD) and terrorism have transformed the role of modern-day forensic science. The term forensic science used to apply only to the scientific analyses of evidence in the context of civil or criminal law. Increasingly, however, forensic analyses are being used to monitor and verify exposure to biological weapons such as anthrax and Sarin, oversee compliance with international treaties and agreements involving weapons of mass destruction, and assist in the investigation of war crimes.

Crime Scene Investigation is not intended as a textbook or a "how-to-conduct-a-death-investigation" primer but rather it will take you on a journey from the crime scene, where you can look over the shoulders of the criminalist, homicide detective, photographer, and pathologist as they identify and collect evidence.

At the morgue, you will explore the role of the forensic pathologist and learn how the victim is identified and the time of death ascertained, as well as the cause and manner of death in suspected murder cases. You will see how evidence—even a small drop of blood or other matter under the fingernail of a victim—can be used to convict a killer. You will discover how the new field of DNA testing is used to catch murderers and how invisible fingerprints are collected from a dead body. You will also be brought into the courtroom to discover how all the different disciplines of crime scene investigation come together to analyze biological and physical evidence to achieve the ultimate goal of the investigative team: the pursuit of truth and justice.

Cyril H. Wecht, MD, JD

01 The Crime Scene

In the investigation of crime scenes, those involving violent death or murder can require the expertise of many more specialists than any other form of crime.

Some crime scenes are found by people going about their normal business. A hiker in the woods may come across a skull; the mailman may notice that several days' worth of mail have piled up in a mailbox; or someone may see a lifeless body in a neighbor's living room. Other times, a gunshot in the middle of the night or a scream for help will alert people nearby that something is wrong. Or an unanswered phone will make a caller check on the person who should have answered. In rare cases, deaths go unreported for a long time—sometimes for years. The bodies, buried in the woods or stashed under the floorboards of houses, only come to light during the laying of underground utilities or home renovation. Or the killer confesses and leads police to the body.

Once a victim has been discovered, the initial reaction is to call the national emergency response phone number to summon medical technicians and the police. This results in the rapid arrival of emergency vehicles with flashing lights and screaming sirens.

First Responders

Among the thousands of calls answered every year by emergency response teams—police officers, paramedics or emergency medical technicians (EMT), or fire crews—there will be a number that involve individuals who are beyond

A DOA—the body of a young man who was declared "dead on arrival" and was covered by paramedics.

medical treatment. These people are called Dead on Arrival (DOA). The diagnosis of DOA must be made cautiously and carefully. Pronouncing a patient dead in the field has both medical and legal ramifications; there are many gruesome but true stories of people waking up in the morgue after being declared dead.

The methods used to determine whether a person is dead vary with the situation. If a victim is found at home or work with no visible traumatic injuries, the first action of the emergency response team is to check for a carotid pulse on the side of the neck for a full 60 seconds. During this procedure, another team member will cut away clothing from the chest so that electrocardiogram (ECG) patches can be applied to test for electrical activity in the heart. If there is no heart activity, the monitor displays a flat line. If the patient has no pulse and isn't breathing (apneic), the medical team checks for rigor mortis (stiffening of the body), particularly at the extremities, and definite dependent lividity (solid purple discoloration due to the pooling of blood in areas of the body closest to the ground). If these conditions are found, the person can be declared dead. In some horrifying cases, determining that a person is dead is much simpler: an individual who has been decapitated or eviscerated, for example, is clearly DOA.

At the moment the person is declared dead by the paramedics, jurisdiction over the body switches briefly from the emergency medical team to the police officers when they arrive at the scene.

Before leaving the scene, the medical personnel assure proper coverage and protection for the corpse. They also write a report naming the technicians involved in the incident, noting the position of the victims or victims, describing any injuries, and listing the actions they have taken.

In major towns and cities, the first police officers to arrive have usually been diverted from local patrol duties, and their role is to secure the crime scene. They will seal the building where a body has been found or encircle an outdoor death scene with a double perimeter of "CRIME SCENE" tape. This prevents unauthorized people from entering the area under investigation. The space between the outer and inner tape perimeters is for personnel not involved directly in "processing" the scene; only personnel actively engaged in collecting evidence are allowed within the inner perimeter.

Failure to preserve the scene adequately may destroy the prosecution's chances of proving a case against the criminal. Although patrol officers usually do not take part in the formal investigation, they know the importance of preserving physical evidence for the specialized crime scene investigators

A crime scene in a suburban area is taped off to protect any evidence the forensic team might find valuable in their investigation.

Reportable deaths

If any one of these criteria has been met, the coroner's representative can swiftly respond and begin to process the crime scene.

- All sudden deaths not caused by readily recognizable disease or where the cause of death cannot be properly certified by a physician on the basis of prior (recent) medical attendance.
- All deaths occurring under suspicious circumstances, including but not limited to those in which alcohol, drugs, or other toxic substances may have had a direct bearing on the outcome.
- All deaths occurring as a result of violence or trauma, whether apparently homicidal, suicidal, or accidental, including but not limited to those due to mechanical, thermal, chemical, electrical or radiation injuries; drowning; cave-ins; regardless of the time between the injury and death.
- Any stillbirth or infant death occurring within 24 hours of birth where the mother has not been under the care of a physician, or where the mother has suffered trauma at the hand of another person.
- Abortions, where considered criminal, regardless of the gestational age of the fetus.
- All hospital deaths that occur as a result of accidental injury during diagnostic or therapeutic procedures, including but not limited to surgical procedures; and all deaths following the accidental administration of excessive amounts of a drug, including but not limited to blood or blood products. In addition, all operative, peri-operative, and post-operative deaths in which the death is not readily and clearly explainable on the basis of prior disease.
- Deaths of all persons while in legal detention, jails, or police custody, including any prisoner who is a patient in a hospital, regardless of the duration of hospital confinement.
- Death due to disease, injury, or toxic agent that occurs during active employment.
- Any death where the body is unidentified or unclaimed.
- Any death in which there is uncertainty as to whether it should be reported.

who will follow them. They will make notes of any evidence that might change before the investigators arrive, such as a cold glass of beer on a table in a warm room, or a dry parking space where the surrounding ground is wet from light rain, clearly establishing the very recent departure of a vehicle.

Other responsibilities of the first police officers at the scene include recording the time of notification; noting who is present at the scene; and if witnesses are identified, separating them from each other.

At this point, it's still only a death scene. The death could be from natural causes, or there could be a more sinister reason. The next step is for the police officers to report the death to the coroner's office.

The Death Call

The death call is directed to the investigative division of the coroner's office. This department is staffed by trained death investigators (DIs), sometimes called deputy coroners, and is manned around the clock. The death investigator who takes the call will determine if the death falls under one of the

Case file:

Charles Manson

On August 10, 1969, officers from the Los Angeles Police Department were summoned to the home of famous film director Roman Polanski and his actress wife, Sharon Tate. They found the bloodied bodies of three men and two women, one of them Tate. Apart from one man, all the victims had multiple stab wounds; all the men had been shot. On the front door, the word "PIG" had been daubed in blood. Investigators also discovered a broken pistol grip and an unusual clasp knife.

Robbery clearly had not been the motive for the crime, since no valuable items were stolen, and detectives concluded that the bloodbath had resulted from a drug deal that had gone wrong.

categories of reportable deaths (see page 19) that would bring it within the coroner's jurisdiction.

Not every death constitutes a coroner's case. In the majority of jurisdictions, the coroner's office is interested only in cases that involve sudden, unexpected, unexplained, traumatic, or medically unattended deaths.

Death Investigators

When a dead body falls under the jurisdiction of the coroner, an investigative team is sent to the crime scene to collect evidence and the body. The team usually consists of two death investigators (DIs), a scene forensic photographer, and, depending on the type of case, a criminalist who specializes in ballistics, fingerprints, or blood splatter.

The DIs conduct the initial phase of the investigation of each death and coordinate with the forensic pathologist throughout the investigative process. DIs receive specialized training in areas such as evaluation of postmortem physiological changes, traumatic injuries, and medicolegal investigation. Interviews with witnesses and relatives,

Polanski, who was in Europe at the time of the killings, was devastated.

Similarities were drawn with two other recent murder cases, which also involved multiple stabbings and words written in blood at the scene. The L.A. Sheriff's Office already had a man and a woman in custody for one of these crimes, Bobby Beausoleil and Susan Atkins, who were both members of a hippy group headed by Charles Manson. While awaiting trial, Atkins had told other prisoners that she and members of Manson's group had carried out the murders at Polanski's house, and that they had plans for undertaking further high-profile killings.

Meanwhile, a long-barreled revolver with a broken grip had been found and handed to the police. Members of Manson's group confirmed that it had been used at the ranch where they lived, and bullets were discovered there that matched those recovered from the bodies.

It turned out that the murders had been part of Manson's plan to spark a race war that he hoped would lead to a breakdown in society, allowing his group to take over. Although Manson was present at only one of the murder scenes, the prosecution convinced the jury that all the crimes had been carried out on his orders. He and eight of his followers are serving life sentences.

Five key crime scene questions

1. Did the death take place where the body was discovered or was it moved?
2. Was there any attempt to alter the scene?
3. Does the scene suggest a motive for murder (robbery, drugs, insurance fraud, for example)?
4. Is a cause of death clearly apparent?
5. Are there sufficient clues (physical evidence) to indicate how the crime occurred, and where the victim and the perpetrator were in relation to each other during the commission of the crime?

photographs and sketches of the scene, investigative data from the police, and other pertinent pieces of information are collected at this stage. The material is documented within the initial "circumstance report of death" prepared by the investigative division and filed at the coroner's office for future reference.

In homicide and criminal negligence cases, the investigative division performs a number of additional duties. These include coordination with law enforcement agencies, conducting arraignments of suspects, preparing warrants and subpoenas issued by the coroner, serving subpoenas on witnesses, and testifying at inquests and court proceedings. The investigative division, together with the administrative division, is also responsible for issuing the death certificate in all cases that fall within the jurisdiction of the coroner. The original death certificate is sent to the vital statistics authorities. Thereafter, certified copies are made available for interested parties.

After the death investigators, the police homicide investigation team, normally consisting of two homicide detectives, is next to arrive at the crime scene.

The Investigation

The processing of a death scene requires the application of scientific methodology to the reconstruction of events, criminal or otherwise. The accurate assessment of a crime scene, however, may only be achieved through the correct interpretation of physical facts and their accurate reconstruction in proper sequence. Because some physical evidence is fragile, fleeting, and easily destroyed (such as

A handgun and blood splatter found in the utility sink at a crime scene.

footprints in snow), it is crucial that those who arrive first at the scene are aware of the significance of such physical evidence and are competent to make appropriate decisions about its handling, identification, and preservation.

In many cases, forensic scientists are able to determine from the circumstances at the scene the most likely sequence of events, which in turn enables them to reconstruct the elements of the crime scene. Needless to say, accurate measurements, good photographs, careful recovery of physical evidence, and competent examination, evaluation, and correlation of all the attendant circumstances are vital to a successful reconstruction.

It is a general principle of law that where evidence is purely circumstantial, it must be conclusive, and it must constitute a reasonable and moral certainty that the accused—and only the accused—committed the crime. Very often, reconstruction of a single fact may be the deciding element in a criminal case. It is therefore crucial that investigators always remember the fact that their purpose is to establish whether a crime has occurred, and if so, to provide the evidence in a court of law for a judge or jury to determine the guilt or innocence of the accused.

A Day in the Life: Death Investigator

Joseph T. Dominick, Chief Deputy Coroner, Pennsylvania

The 7:00 A.M.–3:00 P.M. shift is the busiest of the day. I arrive at the office at 6:45 A.M. to receive a report from the investigators completing the night shift. Today, they inform me that there is a body to be retrieved from a local hospital. This is the case of a 35-year-old male who was found unresponsive in his bathroom. He is a known drug abuser but has no other significant medical history.

The remaining three daytime investigators drift in, and I apprise them of the hospital case. Then comes a call about an apparent suicide. My partner and I head to our ambulance and depart for the scene. Generally, we work in teams of two so that nothing is missed and appropriate witnesses are on hand when establishing a chain of custody for any evidence collected.

On arrival, we are met by two police officers and a homicide detective. They explain that the victim is a 22-year-old female, who has been depressed over the loss of her job, financial problems, and a recent breakup with her boyfriend. We establish some vital statistical information and are introduced to her parents. We explain what we will be doing and confirm details of the deceased's history. We also inquire about funeral arrangements.

My partner and I follow the officers to the woman's apartment, meeting the photographer and other police officers securing the residence. We establish that photos have been taken and that the scene has been initially processed. We perform a cursory examination of the body. There is a circular injury entering the right temple and exiting at the left temple, with powder residue around the rim of the entrance wound. A pistol is in the victim's right hand. After disarming this and recording its serial number, we give it to the detective for submission to the crime laboratory. We make a positive ID by comparing the deceased to the picture on her driver's license. Photos are taken of the body, which then is searched for effects. A chain of custody is initiated, and a receipt for the items is signed by the detective.

The body is laid on a clean sheet, and the hands are covered with paper bags. We record the liver and ambient temperatures. Finally, the corpse is put into a body bag and taken to the ambulance. We answer the family's questions about what has occurred and what will happen, then we head back to the office.

On our return, we log in the corpse, obtain a weight, and take an identification photo. The body is placed in the cooler to await autopsy, and the personal effects are submitted to our supervisor. We enter the case on file, having satisfied ourselves that the evidence indicates a suicide. However, the final decision rests with the pathologist. Often our involvement ends here, but sometimes we are called to court to testify about the crime scene or chain of custody.

My department of 14 investigators handles an average of 8,500 cases a year.

The location of the crime is not always simple

At the beginning, the investigation of any death scene is much like attempting a crossword puzzle without the clues. As information is collected and evaluated, however, the clues start to emerge, one clue leads to another, and frequently the puzzle can be solved.

For many people, a crime scene conjures up the vision of a body found after a terrible traumatic event. Classic examples include a tragic motor vehicle incident where the victim is discovered behind the wheel of a mangled car; the residence where an exceptionally depressed person has blown out his skull with a shotgun; or perhaps the street corner where a young man is lying with multiple gunshot wounds to his body. Certainly, these are all settings where an exhaustive and meticulous investigation is essential to get to the underlying cause and manner of death. In many instances, however, there may be several locations that can be considered crime scenes, and each must be examined carefully as quickly as possible before any useful evidence disappears.

Also, in today's high-tech medical environment, the victim from a crime scene may be transported by ambulance or helicopter to the nearest medical facility. There, the person may be stabilized, then transported to other facilities where he or she may remain for days, weeks, or even years before succumbing to injuries. Such complicated cases also require investigation by the coroner.

Once an individual is declared dead by a hospital physician, the body falls under the jurisdiction of the coroner, notification will be made, and the immediate scene will be secured. All intravenous lines and fluids, endotracheal tubes, and other invasive medical paraphernalia must be left in place. Medical records will be requested for review by the pathologist responsible for the case. In many instances, a subpoena will be required to obtain such confidential medical documents.

In the most common situation, where the victim is dead and has not been moved from the site of the incident, the death investigator takes charge of the investigation. Even though he or she has authority over the body and its surroundings, the investigation will be conducted by several specialists, each of whom may bring years of experience to the case.

The victims of a multi-pedestrian motor vehicle accident have been covered by the police. The vehicle came to rest after hitting a tree.

It is essential that all death investigations be handled with a team approach. If anyone adopts the attitude that "the scene is mine and I am in charge," the work usually becomes counterproductive; such a state of mind really has no place in the investigation. Everyone involved has some specific expertise to contribute, which may be essential in solving the case. Ignoring this need for cooperation can quickly result in a botched investigation.

On-Scene Procedures

Before entering the actual scene, the death investigation team will seek basic information from the first responders—who is dead and what has happened? At the same time, the forensic photographer will take overall photos of the scene. Police officers, medical professionals, and witnesses on the scene will be interviewed by the DI to get an understanding of possible scenarios or preliminary theories that may be important, and to ascertain the facts that have been established. Obtaining vital statistical information such as a tentative name, address, age, date of birth, social security number, and phone numbers of the victim is a priority, as is next-of-kin information, if it is available. Past medical, psychological and social history, medications regimen, and any other significant data about the victim should be established. An accurate record of the name, title, phone number, and affiliation of everyone at the scene is also essential.

Moving from the outside inward

Once the overall photographs have been taken, the investigation moves to the scene itself. The DI will examine the scene from a distance initially, then move progressively closer until he or she arrives at the victim, all the while making and recording observations of the scene. When the DI finally reaches the deceased, his or her attention will switch to establishing information about the corpse. The

The police search for clues following Washington D.C.-area shootings. A 37-year-old man was shot in October 2002 as he left the Ponderosa restaurant with his wife.

investigator will make a general check of the body, noting any obvious injuries, and carefully checking its position and condition. Copious notes and photographs will be taken, showing whether the corpse is in a supine, prone, or lateral position, and close attention will be paid to the state of the body's rigor mortis. Rigor mortis suggests to the investigator an approximate time of death and whether the body's position has been altered since death.

Examination of the body

Then the investigator's attention will be directed toward the clothing and other items of significance on or around the body. The corpse will be searched thoroughly by the DI, in the presence of police officers, to retrieve any relevant effects. Any item discovered will be inventoried, recorded, and, if necessary, photographed. A chain of custody for all of the effects and evidence gathered must be initiated by the DI at this time. Finally, the evidence and effects will be secured by the DI for submission to the crime laboratory as evidence, or for return to the next of kin.

Case file:

Frederick and Rosemary West

In the summer of 1991 Frederick West from Gloucester, England, was facing a charge of sexual assault against several of his children. The case was dropped, however, when his daughter refused to give evidence against him. Later she told a police officer that she was frightened of ending up in the backyard like her sister. The sister had disappeared four years before. Alarmed, the police arrested West and his wife, Rosemary, and started a painstaking search of their home.

Fred West soon admitted to murdering the girl and described exactly where she was buried.

Removing the Body

The body itself is a major piece of evidence, and once the DI has completed the investigative work at the scene, it can be transported to the morgue, where a postmortem examination will be performed. Before it can be moved, however, certain actions must be taken.

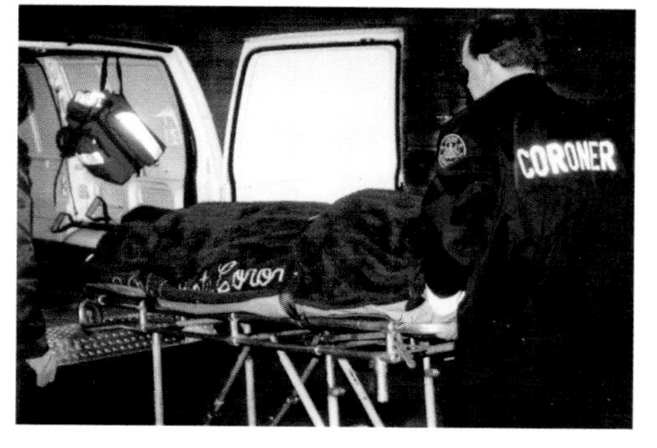

A death investigator prepares to transport a body in the coroner's van.

Positive identification

Even though a tentative identification may have been made prior to the investigator's arrival, if at all possible, a positive ID should be carried out at the scene. Some form of photographic identification, such as a driver's license, would be the first choice; otherwise, someone on the scene who knows the deceased personally might make a visual identification. If no positive ID can be made, the remains will be listed as unknown until some more accurate method, such as fingerprints or dental records, can be employed.

Hands and weapons treated as crucial evidence

In the event the death was due to a physical altercation or involved a gun, the hands of the deceased must be placed in fresh paper bags and secured with tape at the forearms. The bags will contain and preserve any evidence that may be on the victim's hands. This might include gunshot residue or trace evidence under the fingernails. It is essential to use paper rather than plastic because paper allows the evidence to "breathe," preventing condensation that can destroy that evidence.

Weapons found at the scene are normally collected by the homicide detectives and submitted to the crime lab for latent fingerprints and ballistic analyses. No attempt is ever made to remove bullets or clothing from the body, or to otherwise change any relationship between the body and its environment while at the scene. To do so would risk destruction or loss of important evidence. Instead, these items will be recovered at the autopsy, where the environment can be controlled.

Moving to the morgue

The body is prepared for transportation to the morgue by wrapping it in a fresh white sheet or shroud, then placing it in

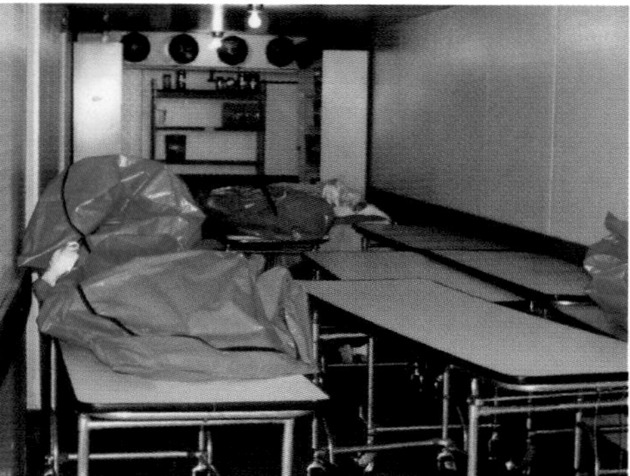
Inside the refrigerated storage cooler at the morgue, several bodies await a postmortem examination.

a disaster or body bag tagged with the appropriate labels. The bag is locked to insure that no tampering takes place en route. The body is lifted onto a stretcher, wheeled to the coroner's vehicle, and driven to the morgue. It is important that the body be kept in its original position while being transported.

Once the victim has been removed from the scene, the police and crime lab personnel will continue to process the vicinity for additional evidence that might be recovered. Adequate time and personnel must be devoted to this process, since there may never be another opportunity to examine the scene or the evidence under the same conditions.

Enter the Specialized Forensics Experts

Depending on the type of crime scene, the skills of a variety of specialized personnel, called criminalists, may be required to assist in the investigation. Criminalists are trained in particular disciplines within forensic science and deal with physical evidence such as hair, fiber, glass, fingerprints, blood splatter, and firearms (see chapter 3). They can help identify, locate, and collect pieces of evidence that may provide additional clues in the attempt to bring the criminal to justice. If necessary, a forensic pathologist, forensic anthropologist, or forensic entomologist may be called in to contribute expertise to the investigation.

Morgue Procedures

Upon arrival at the morgue, the body's weight is recorded. The corpse is photographed again and logged in to the autopsy suite, where it is kept in a refrigerated storage vault until the pathologist and his team are able to perform their examination. All necessary paperwork is completed by the DI, and a written report is entered into the coroner's computer network.

All evidence and personal effects (such as a wallet and credit cards) recovered at the scene must be logged in and turned over to the appropriate department to maintain the chain of evidence and custody. Any follow-up calls regarding the investigation will be made at this time, and any subpoenas for medical and police records will be issued. The investigator must then make every effort to notify the next of kin.

A Day in the Life: Forensic Photographer

Marty Coyne, Chief Forensic Photographer, Pennsylvania

At 4:00 A.M., the telephone jangles me awake and gets me out of bed. It is the 911 dispatcher, telling me that I am needed at a death scene; a body has been found in a vehicle parked near the river. The outside temperature is in the single digits, so warm clothes are essential; I could be outside for several hours. This is a 24-hour, seven-day-a-week job.

When I arrive, the detectives are already there. The passenger-side window has been shattered, and the individual in the driver's seat has what appears to be a gunshot wound to the head. I start taking photographs, on my own initiative and at the request of the scene investigators. First, I document the outside of the vehicle, then the shattered window and the glass fragments inside, followed by the victim seated in the vehicle. I take care not to miss the small things, such as anything in or near the hands and whether the feet are on any of the pedals. I pay particular attention to the wound, and any blood trails and drops on the body.

Once the body has been removed, I take a full-face ID photograph, along with photos of anything on the body or clothing that may be of value or interest. Then, once again, attention is turned to the vehicle and surroundings. The positions of the transmission lever and the ignition key are documented, as are any items on the seats. If there are any shoe impressions nearby, they will be photographed, as well as the contents of trash cans and possibly even the local storm drain. In some instances, more images may be needed of the vehicle when it is examined at the forensic lab.

Now it is time to head for the coroner's office to document the autopsy being performed on the victim.

The body, dressed just as it was when removed from the vehicle, is placed on the autopsy table, and overall views are taken. The forensic pathologist and autopsy technicians go over the corpse, describing the body and clothing. Findings that may become important are photographed as they come to light.

Then the body is undressed and the process repeated. After the body has been washed, photos are taken of any injuries, along with the damage caused by the gunshot. The hands are photographed to show the presence or absence of defense wounds. Damage to any organ is photographed, usually "as found" and again on a special stand.

Artifacts such as bullets are also recorded "in situ" or in their natural position and on the photo stand. The range of images for any item is always the same: an overall photo to indicate the location on the body, a mid-range view to pinpoint the location, and a close-up to show the detail.

The exposed film goes to our own photo lab; it is essential that we keep custody of this valuable evidence. Occasionally, I am called to court to testify to that fact.

The mission of the coroner

The coroner's mission is threefold:

1. Determine the cause of death.
2. Determine the manner of death.
3. Determine the mechanism of death.

At this point, responsibility for the death investigation shifts to the pathologist, toxicologist, and criminalist. The type of examination will be determined by the forensic pathologist in the light of the circumstances of death.

The coroner and the classifications of death

The cause of death is the pathological condition that produced the victim's death. Typical examples include arteriosclerotic cardiovascular disease, pneumonia, pulmonary embolism, gunshot wound, blunt-force trauma, and drug overdose.

The manner of death is a subjective or interpretive opinion that explains how the cause of death came about. It is a legal interpretation that the coroner assigns to a particular case. In general, there are four classifications for manner of death: natural, accident, suicide, and homicide. Occasionally, the coroner will list the manner of death as pending or undetermined. The former simply means that further testing or investigation is warranted before a decision can be made; the latter is used when all avenues of investigation have been explored, yet a valid conclusion cannot be reached. A good example would be a case in which the victim was playing Russian roulette and shot himself in the head, thus causing his death. Was this case a suicide, an accident, or possibly even a homicide? Many theories might exist, but no one knows for certain what the deceased was thinking prior to the shot penetrating his head. Therefore, classifying the manner as undetermined would be appropriate.

The mechanism of death refers to the physiological changes that took place within the victim leading to death. Examples include cardiac arrhythmia (irregular heartbeats), hemorrhage, and anoxia (complete lack of oxygen in the body). It is important to understand that the same mechanism may be associated with a variety of causes of death. For instance, hemorrhage may occur because of a gunshot wound or the rupture of a blood vessel that has been eroded by cancer.

If the death is considered natural, no criminal investigation is needed, although a thorough examination of the facts leading to the death must be undertaken. A ruling of accidental death may result in a continuation of the investigation, but in a different direction. There may be civil liability, and the coroner may be called to testify in civil court. The amount of time devoted to this extended investigation will vary from one jurisdiction to another.

When the manner of death is given as suicide, this will need thorough confirmation. In most cases, the determination will stand; occasionally, however, new evidence revealed by an in-depth investigation will change the coroner's mind.

Homicide is the taking of a human life by another human being. In most instances, it involves a criminal act, but homicide does not necessarily indicate criminality in every case. A homicide may be considered justified, as in the case of a police officer returning fire, and killing the attacker. It may also be excusable, for example, when someone is killed in an act of self-defense. Under all circumstances of a homicide ruling, a comprehensive and painstaking investigation is carried out by the coroner and his or her office.

The investigation usually does not end when the autopsy report has been completed. The coroner or, more often, the forensic pathologist who conducted the postmortem examination will be called to testify on his or her findings during the criminal proceedings. In some cases, such as fatal child abuse or poisoning, the report forms the major component of the prosecution.

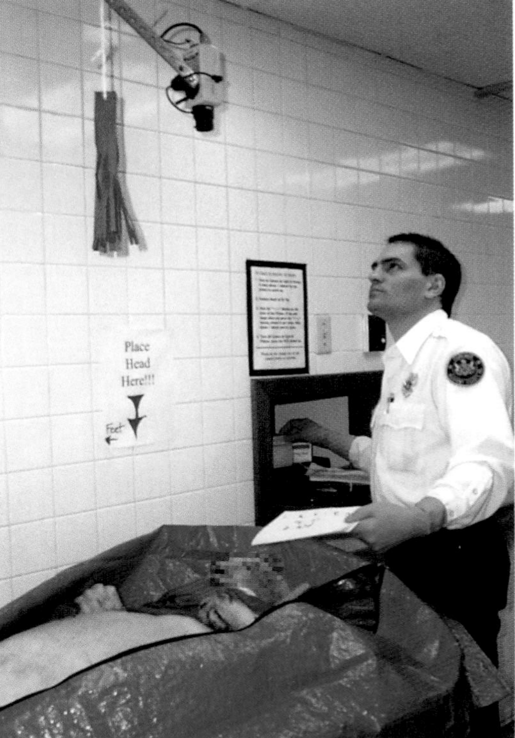

A death investigator takes an ID photograph of a newly arrived body at the morgue.

02 The Inquiry Team

The mission of the criminal justice system is to deliver justice. To achieve this goal, a homicide investigation has to determine that a crime has been committed, identify and apprehend the killer or killers, and present the evidence in a court of law.

Soon after a victim's body has been found, declared dead, and the scene secured, the emphasis turns to collecting physical evidence. This evidence is vitally important as it can show that a crime was committed, place a suspect at the scene when the crime was committed, prove the suspect is the killer, and exonerate an innocent person.

All members of an investigative team know that the individual pieces of evidence they collect may be used in a scientific reconstruction of the murder. The accurate

Principles behind the gathering of physical evidence

1. Every contact between two people leaves trace evidence on both of them.
2. Trace evidence that can be found, documented, and examined can link a particular individual to a specific place and/or time.
3. Trace evidence can be identified by class characteristics, individual characteristics, physical matches, and mathematical probability.

The body of a woman, found in April 2003 at Point Isabel on San Francisco Bay in California, is taken to the coroner to be identified. Police later identified it as that of Laci Peterson, who disappeared mysteriously in her eighth month of pregnancy.

reconstruction of a crime scene can only be achieved through the correct interpretation of the physical facts, followed by their reassembly in proper sequence.

Typically, after a victim has been declared dead, two homicide detectives and two DIs arrive at the scene. Both teams conduct an initial survey of the death scene to determine what type of criminalist needs to be called to the scene.

The Homicide Detectives

The detectives who arrive at a crime scene will seek answers to many of the same questions that concern the other members of the investigation team. These include:

- Who was killed?
- How, where, and when were they killed?
- Why were they killed?
- By whom were they killed?

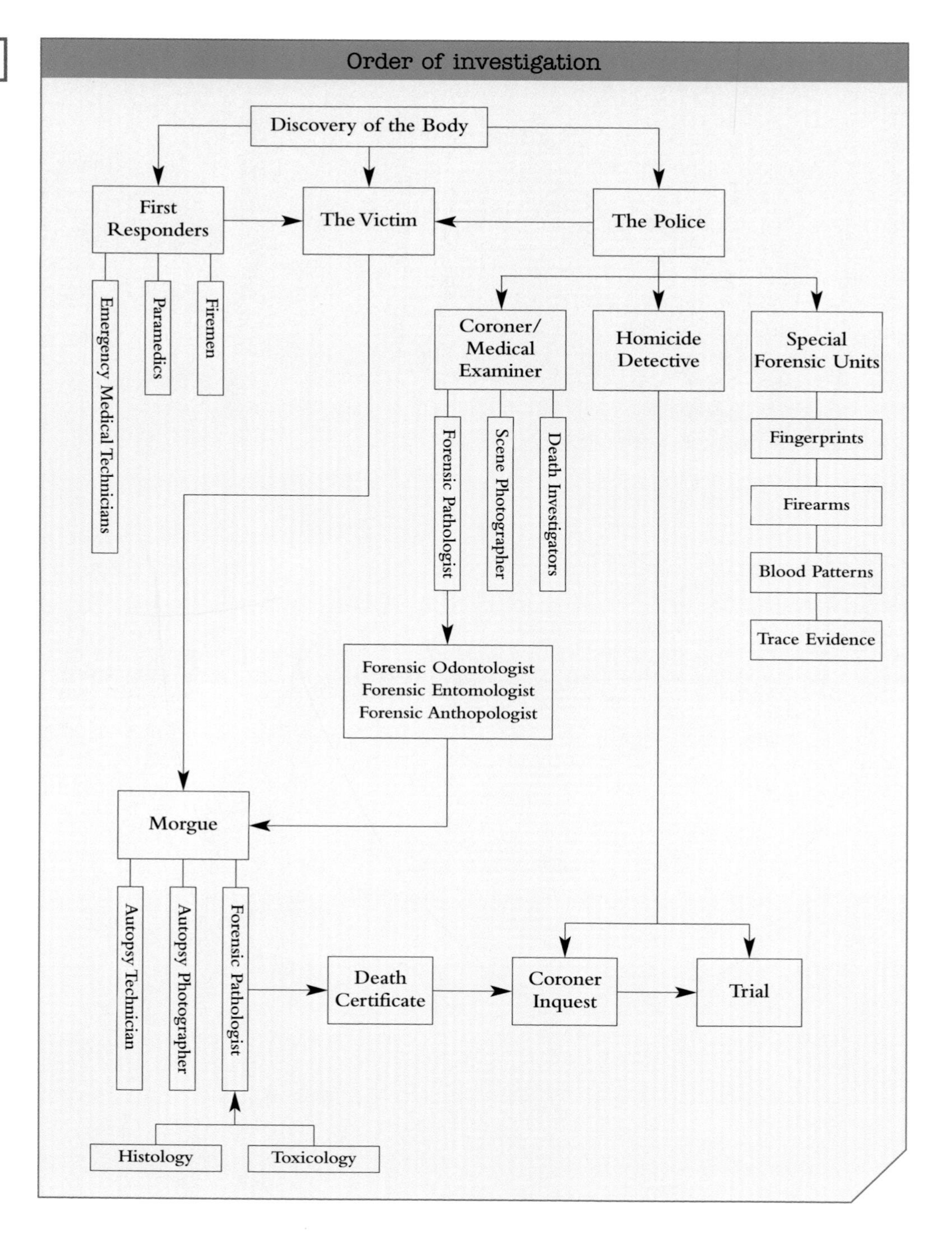
Order of investigation
Discovery of the Body
First Responders
The Victim
The Police
Emergency Medical Technicians
Paramedics
Firemen
Coroner/ Medical Examiner
Homicide Detective
Special Forensic Units
Forensic Pathologist
Scene Photographer
Death Investigators
Fingerprints
Firearms
Blood Patterns
Trace Evidence
Forensic Odontologist
Forensic Entomologist
Forensic Anthopologist
Morgue
Autopsy Technician
Autopsy Photographer
Forensic Pathologist
Death Certificate
Coroner Inquest
Trial
Histology
Toxicology

The main role of the forensic pathologist is to answer the "Who," "How," "Where," and "When" questions of the investigation during the subsequent autopsy. After a murder suspect has been identified, the criminalists can match him or her to a crime by trace evidence. Often, fingerprints at the scene will suggest a killer. However, the pathologist and criminalists cannot answer the two remaining questions: "Why was the person killed?" and "Who committed the killing?" Finding these answers falls to the detectives. They collect evidence by interviewing the victim's family, friends, and co-workers; and they review the victim's employment history and criminal record. Detectives also study the victim's bank and credit-card activity. As they assemble information, the detectives begin to see possible motives for the killing. Suspects begin to emerge. Each suspect is compared to the physical evidence and either dismissed or subjected to further investigation.

Police inspect a murder crime scene, in Balham, London. Special units including medical examiners, homicide detectives, and criminalists will all become involved in this case.

Forensic Photographers

A forensic photographer, like other members of the death investigation team, has an intricate part to play in the death inquiry. He or she must be ready 24 hours a day, seven days a week, to go anywhere a death has occurred. Moreover, the forensic photographer is not a passive member of the death investigation team. His or her experience, insight, and ability to document evidence are invaluable to the medicolegal system.

The primary challenge of a forensic photographer is to discover all the details of the case, then to provide an accurate and complete photographic record of those details. These photographers treat every scene as a crime scene and document it that way. Later, the images act as powerful pieces of evidence in court. The most natural, innocent-looking death may become a sinister case, while the most grisly-appearing death may turn out to be from natural causes. Without the photographic details provided to the homicide investigation by the forensic photographer, such crucial determinations may not be made correctly.

Case file:

Malcolm Fairley

In 1984, a series of violent burglaries took place in southern England. The offender would threaten his victims with a sawed-off shotgun, rape the women, and beat the men. He always wore a hood that covered his face, and the only clues detectives had were that he had a northern accent and was left-handed. A massive police hunt ensued, but with so little to go on, the investigators made no significant progress. Meanwhile, the attacks received regular exposure in the newspapers, which dubbed the man "The Fox."

On August 17, he struck again. As it turned out, this attack was opportunistic, since "The Fox" had been on his way north to visit his mother and had decided on a whim to carry out the burglary. It would be a serious mistake. He parked his car out of sight in a field near the village of Brampton, fashioned a makeshift hood from some coveralls he had with him, grabbed his shotgun and broke into a nearby house. There he bound the male resident and raped the man's wife, after which he calmly removed any obvious evidence that could tie him to the crime, including part of a bedsheet.

When police investigators arrived at the scene, they discovered tire tracks where the car had been parked, as well as a paint chip, which had been dislodged when the car had rubbed against a tree. Nearby was the section

There are three types of forensic photographer: the scene photographer, the technical photographer, and the autopsy photographer.

Scene photographer

As the name implies, the scene photographer attends the death scene, wherever it may be. Usually, it is indoors. After locating the body, the photographer will use the "four corners" method to record an indoor scene. First, a series of photographs is taken from the doorway of the room in which the victim was found. Then the photographer moves to the three remaining corners of the room in turn, producing a panoramic view of the scene. Next, he or she concentrates on the victim, taking full-length photographs of the body, views from the left and right sides, and close-ups where needed. Once the body has been photographed in detail, the photographer documents the

A homicide scene. The yellow numbered markers identify forensic evidence ready for expert examination.

of bedsheet, the hood, and the shotgun, which had been hidden under some leaves. Certain that the offender would return to retrieve his gun, the police mounted a surveillance operation. To account for their presence in the quiet country lane, they faked a car accident, but he never showed up.

Meanwhile, the paint chip had been analyzed and turned out to be a color called Harvest Gold, which had been applied to one make and model of car, the Austin Allegro, for two years only.

With nothing else to go on, detectives began tracing and interviewing all the burglars known to have moved south from the north of England—over 3,000 of them. In the course of this search, two police officers went to the home of Malcolm Fairley, where they found him outside washing his car, a Harvest Gold Austin Allegro. Fairley spoke with a northern accent and appeared shifty when they questioned him; when he reached into the car to pick up his wristwatch and put it on, the officers noticed that he was left-handed. His car was damaged in a place that matched the position of the paint chip found at Brampton. He was arrested, and a subsequent search of his apartment revealed identical coveralls to those from which the hood had been cut. He confessed and was given six life sentences for his crimes.

When long periods of time pass between a crime and its hearing in court, the forensic photographer's role extends to creating accurate diagrams and illustrations of both the crime scene and the body.

surrounding area, including any potential weapons, spilled drinks, and ashtrays with their contents. All rooms directly connected to the room in which the body was discovered must also be photographed panoramically. Anything unusual within the adjoining rooms will be documented.

It can be challenging to differentiate between the usual and the unusual. Most people, for example, would not consider a newspaper tossed casually into a trash can to be unusual. If, however, the victim was an advocate of recycling or someone who was not known to read that particular newspaper, such a sight might be considered unusual. Smashed glassware on the kitchen floor, a broken door or window lock, or a disrupted dresser are generally considered unusual and will be photographed. Once the interior of a crime scene has been documented, exterior features such as pathways, balconies, and fire escapes are photographed, as are any signs of forced entry, such as broken windows, torn screens, and tool scrapes, even if they appear to be old.

Photographing auto crashes

The second most common death scene covered by forensic scene photographers is auto crashes. One vehicle may strike a person, an object, or another vehicle. In all these cases, an overall view of the scene will be taken first, followed by photographs of any victims inside or outside the vehicle; the relationship between the location of the victim or victims and the vehicle; the interior, exterior, and mechanical conditions of the vehicle or vehicles involved; and the landmarks surrounding the scene, such as utility poles, signs, and traffic control devices. The photographer must also document any drag or skid marks.

Victims will be photographed to show any patterns seen on the clothing or the body that may have been caused by parts of the vehicle, such as the tires, grille, license plate, or headlight covers.

Overall and close-up photographs of the vehicle must be taken, including damage (old and new), any stickers that indicate its last safety inspection, and any area that may have evidence attached, such as fibers from clothing, blood smears, hair, or tissue.

Scenes where vehicles have struck objects can be simple or quite complex. A simple scene would be one car, occupied only by the fully restrained driver, hitting a solid, fixed object, such as a telegraph pole. A complex scene could involve a vehicle with several occupants, some of whom may not have

At a scene of a two-vehicle accident, the victim has been covered by paramedics.

been restrained and even ejected, striking several objects, with no indication of which person was driving.

Photographs should establish the general location of the vehicle or vehicles, occupants, and pedestrians at the scene, including the road leading to the impact site. They should record scrapes or paint on barriers and tire marks. Within the vehicle, the victim or victims, instrument panel, transmission lever or gearshift position, position of the ignition key, control settings (such as the headlights, windshield wipers, and heater/air conditioning), the restraint belts, and the deployment of the airbags should all be photographed. If there is a question about which occupant was driving, photographs may assist the pathologist in making that determination. Airbags deploying from the steering column produce a different pattern of propellant than the dashboard safety bags, and side airbags will leave their patterns on opposite sides of the passenger's and driver's upper torso and head.

In cases of apparent single-vehicle, single-occupant fatal accidents, one may ask why send a photographer? First, it may

Case file:

Byron de la Beckwith

In the early hours of June 12, 1963, Medgar Evers was shot in the back outside his home in Jackson, Mississippi; he died in front of his wife and children. Evers had plenty of enemies. As a field secretary for the National Association for the Advancement of Colored People (NAACP) and agitator for racial equality, he had organized sit-ins for equal access to public accommodations and worked on major voter registration campaigns. He had provoked the ire of local white supremacists.

When police arrived at the Evers home, they discovered a rifle that was registered to Byron de la Beckwith, a member of the Ku Klux Klan. Examination of the weapon produced his

The victim of this fatal motorcycle accident, located under the sheet, lost control of the vehicle. The body ended up being separated from the motorcycle by 15 ft. (4.5 m).

fingerprints. He was arrested and tried in the spring of 1964. However, two police officers testified that they had seen him some 60 miles away at the time of the shooting. The trial collapsed with a hung jury; a subsequent trial in the fall of that year produced the same result.

However, Evers' widow never stopped pushing for a retrial. Finally, in 1989, papers came to light that indicated the possibility of jury tampering and official misconduct during the second trial. As a result, Assistant District Attorney Bobby DeLaughter accepted the case, despite the passage of years, lost evidence, and the fact that many witnesses had since died.

DeLaughter got a break when the prosecution discovered a book about the activities of Delmar Dennis, a former KKK member turned FBI informant. At a Klan rally, Dennis had overheard de la Beckwith boast about shooting Evers. The murder weapon was subsequently found in a gun collection belonging to de la Beckwith's father-in-law. Evers' body was exhumed, allowing the fatal wound to be compared with bullets fired by the rifle.

In 1994, de la Beckwith, by then 73, was tried again for the murder of Evers. This time he was found guilty and sentenced to life imprisonment. He died from heart problems while in jail in 2001.

not be a single-vehicle incident. There have been countless instances where one vehicle, accidentally or intentionally, has forced another off the road, then fled the scene. Second, it has been estimated that up to 40 percent of all single-vehicle, single-occupant collisions with fixed objects may be suicides, not accidents. This is particularly likely if the road, vehicle, and weather conditions are good. There are no hallmarks that absolutely differentiate these suicide-by-motor-vehicle incidents from genuine accidents, but the presence or absence of certain indicators can point toward the correct conclusion.

It is the responsibility of the photographer to capture those indicators: marks on footwear and the pedals of the vehicle, for example; the vehicle's lights are turned off, even though it was being driven at night; the windshield wipers are off, even though rain or snow was falling at the time. Finding a driver's hands tucked under the lap portion of the seat belt suggests a suicide rather than an accident. The object or objects struck—trees, abutments, utility poles, or buildings—must be fully documented.

Suicide must be ruled out in a single car fatality and photographic evidence, such as how the victim's body is found, can be invaluable to investigators.

The procedures for photographing accidents involving more than one vehicle are essentially the same as those for a single-vehicle accident. There are, however, additional considerations: the chain-reaction phenomenon, for example. Deaths can be scattered along the length of a chain of cars or trucks in a pileup. Individuals may have been ejected from one vehicle and subsequently struck by one or more others. In rare instances, police, fire, and rescue personnel may become collateral victims, struck while trying to help the first victims. Each vehicle involved in a chain incident must be photographed completely, even those with only minor damage or in which no occupants were injured or killed. The most heavily damaged vehicles, however, usually yield the most information.

Documenting workplace mishaps

The third type of death scene encountered by forensic photographers is an industrial or workplace site. An industrial death occurs when someone is killed performing an action related to his or her occupation. Typical locations include mines, construction sites, retail outlets, warehouses, power plants, airfields, and offices. The law requires the investigation of such deaths. From the photographer's point of view, however, a work-related death is documented in the same manner as a homicide scene. Evidence can include almost anything from a frayed power cord on a PC to a steel girder, an improperly parked vehicle, or a wood retaining wall in a ditch.

One of the most complex and dangerous jobs that a forensic photographer faces is documenting a fire scene. These are generally poorly illuminated, waterlogged, and likely to have pockets of noxious, toxic or flammable gas, and other dangers that can harm the photographic equipment and the photographer. The fire's cause must be documented, along with variables such as improper building or renovation procedures, missing or non-operative smoke detectors, illegal storage of chemicals or other flammable materials. Sometimes the activity that led to the fire becomes more important than the fire itself and its consequences, producing two sets of requirements for photographic documentation. The location and position of a body at a fire scene may be of great importance in determining not only the cause and manner of the fire, but also the cause and manner of the death.

Everything that could be linked to a fire must be photographed. Such items include light bulbs that may indicate the direction of the greatest amount of heat, or the melted solder on a water pipe that might suggest a dry sprinkler system. Dresser drawers that have been tossed about and closets that have been ransacked might show that arson has been used to cover a burglary. An electric clothes iron, set to one of the "on" positions and discovered next to a body in a basement laundry room, might indicate an accident, while the same iron found in the shell of a burned sofa could point to something more sinister. The position of a body may show that the person was attempting to flee the fire. The photographer must capture these situations so that investigators may use them to answer all the questions posed by the death.

Teamwork for mass-casualty disasters

The fourth type of death scene that forensic photographers will encounter is a mass-casualty scene, where a team of photographers is required. The actual death or victim count

Case file: Robert Hansen

On June 13, 1982, a police officer at Anchorage Airport, in Alaska, encountered a distraught young prostitute. She claimed to have been kidnapped at gunpoint, taken to her attacker's home, and raped violently. Then he had told her that he intended on flying her to a backwoods cabin for more sex. While he had been loading the airplane, she had escaped.

The woman's description of her attacker matched that of a local businessman, Robert Hansen, who owned an airplane that he used on hunting trips. Although she identified Hansen's airplane and house, he denied everything, saying that she had attempted to blackmail him. Moreover, two business associates gave him a cast-iron alibi. Nevertheless, the police were suspicious, particularly since they had found the bodies of two women in the wilderness in 1980.

In September 1982, a woman's body was discovered in the remote Knik River region. She had been shot with a high-powered rifle, apparently while naked, since there were no bullet holes in her clothing. A year later, another body was found in the same area in similar circumstances.

Convinced that Hansen was the prime suspect, the police called in a profiler, who suggested that, because of his appearance (Hansen was short, had disfigured skin and

that triggers a mass-casualty plan will vary from one jurisdiction to another; what constitute mass casualties will differ somewhat between a large city, like Los Angeles or New York, and a small rural town. However, a mass-casualty scene is usually a situation that involves more than 10 deaths. The cause can be natural calamities, such as floods, hurricanes, tornados, earthquakes, and landslides; or man-made disasters, such as nightclub fires, aircraft crashes, the release of toxic substances into the air or water, or terrorist attacks.

A victim recovered from this small jet airplane has been placed into a body bag. Red flags can be seen behind the plane. These flags indicate the locations of body parts and sections of the plane.

Covering these traumatic events takes teamwork. After an airplane crash, for example, the site is first marked off and photographers are assigned specific areas to document. Others will photograph recovered effects and parts of the aircraft itself, or work with the forensic autopsy and victim identification team, particularly photographing mandibles and dentures for comparison with dental records.

spoke with a stammer), he could have had trouble with girls as a teenager and that the murders were a means of venting his anger against women. All of the victims had worked either as prostitutes or topless dancers and tended to live a transient lifestyle, which meant that they were the sort of people who were unlikely to have been missed for some time. It was thought that initially Hansen would have used the airplane to dump the bodies where they were unlikely to be found, but later to satisfy a bizarre need he would have begun stripping his victims, releasing them and hunting them down like wild animals. The profiler suggested that since Hansen was a hunter, he was likely to have trophies of the game he had shot in his home; similarly, it was thought that he would have taken trophies from his female victims to remind him of his achievements.

Then Hansen's business associates admitted they had lied, and the police raided his home. They found a concealed Ruger hunting rifle, jewelry and ID cards from the victims, and a map marked with the locations of the bodies. Ballistics tests on bullets fired by the rifle matched them to bullets that had killed two of the women. Hansen confessed to four murders, a rape and a kidnapping; he was jailed for 499 years.

Technical forensic photographer

Some death scenes require the skills of technical forensic photographers. Their specialty is to record bloodstains, blood spattering, and finger, shoe, or tire impressions found at the scene or on the body, using special film and cameras that are capable of taking highly detailed images in scale. Technical photographers spend most of their time working with high-magnification photomacrography, photomicrography, non-visible wavelength imaging, and digital image manipulations.

Each photograph produced by the technical forensic photographer will contain some form of scale to give an accurate idea of the true size of the object being shown. The scale may be something as simple as a 6-in. (15-cm.) ruler, or it may be far more sophisticated. If you ever found yourself at a crime scene with a camera and you noticed that evidence, such as a blood pattern, or tire or shoe tracks in the snow, was being threatened by the weather conditions, you could use any item of known size to establish scale, such as paper money, a coin, or even a credit card.

Autopsy photographer

After a body has been transported to the coroner's facility, a forensic pathologist conducts a postmortem examination. This is recorded by an autopsy photographer, who will be expected to have a basic knowledge of human anatomy. Photographs taken during the autopsy begin with full-length images of the body (front and back) as it arrived at the morgue, followed by the same views after the clothing has been removed and the body cleaned. Close-ups will be taken of wounds, bullet holes, fractures, surgical scars, and other identifiable marks such as tattoos. These photographs are important in establishing the identity of the body undergoing the autopsy, as well as documenting any injuries. Subsequently, such images may aid the identification of an unknown individual. During the autopsy, the organs of the body are photographed twice: first, "in situ" to show the location and severity of disease or damage; second, after the organs have been removed and cleaned.

The value of photography

The role of the forensic photographer does not end after the crime scene and autopsy photographs have been taken. Photography provides a permanent record of events. The crime

scene photographs can serve as a link between the investigator's report and the testimony of others. No one can predict how much time will pass between the date of an incident, the arrest of a suspect, and the actual start of judicial proceedings. Photographs can readily display information that is too small to be seen by the naked eye, such as the cancerous cells of a tumor, or information that is too distant or not normally visible, such as aerial views of a crime scene. Photographs can be used to show things that would otherwise be totally invisible because the human eye is not sensitive to the wavelengths involved, such as infrared and ultraviolet images.

Every piece of evidence must be photographed by the scene photographer and cataloged before it is removed from the crime scene for forensic examination.

Events that occur too quickly for human senses to detect, such as the moment of impact of a bullet, can also be captured on film and demonstrated at a more comprehensible pace. Autopsy photographs can provide a meaningful description of the deceased's condition to jurors. A medical description of the brain injuries suffered by a bludgeoning victim would probably mean little to the average juror. A photograph of the damage makes the trauma quite clear. An investigator's sketch and verbal testimony about the condition of a ransacked room could be misinterpreted by different individuals. A photograph clearly shows what happened there. In addition, photography plays an important role in cases of mass destruction and terrorist actions, allowing the full impact of the event to be understood. When a case comes to court, the forensic photographer is also responsible for creating accurate diagrams and illustrations of both the scene and the body.

The Forensic Pathologist

A forensic pathologist—who works for the coroner—is a physician with specialized medical and forensic science training and knowledge, which provides an understanding of the types and causes of injuries, and the causes of sudden and unnatural death. Forensic pathologists frequently take part in death-scene investigations, perform forensic autopsies, review medical records, interpret toxicology and other laboratory

Case file: Tracie Andrews

One evening in December 1996, police officers were called to a quiet country lane near the English town of Alvechurch, Worcestershire. They found a blood-spattered Tracie Andrews cradling her dead boyfriend, Lee Harvey. He had suffered almost 40 stab wounds. Andrews claimed that he had been the victim of a road-rage attack, and that when she had tried to stop it, the assailant had beaten her off, then driven away.

A couple of days later, Andrews made an emotional appeal in the media for the attacker or any witnesses to come forward. However, the only people to have seen the couple at about the time of the attack had

studies, certify sudden and unnatural deaths, and give court testimony in criminal and civil law proceedings. The scope of their medicolegal investigation addresses the following questions:

1. Who is the deceased?
2. Where did the injuries and ensuing death occur?
3. When did the death and injuries occur?
4. What injuries are present (type, distribution, pattern, cause, and direction)?
5. Which injuries are significant (major vs. minor, true vs. artifactual or postmortem)?
6. Why and how were the injuries produced? What were the mechanisms causing the injuries and the actual manner of causation?
7. What actually caused the death?

Often the forensic pathologist attends the crime scene to begin a scientific reconstruction of the events leading to the death. Visiting the scene is also critical in determining

noticed no other vehicles. A young girl in the vicinity had heard an argument, but only between a man and a woman.

As the police began to make further inquiries, they learned, contrary to Andrews' assertion, that the couple's relationship was somewhat tempestuous, and that once she had attacked him with a bottle. She had also threatened a previous boyfriend with a knife.

At this stage, the forensic evidence began to make a major contribution to the investigation. The pattern of bloodstains found on Andrews' clothing and at the scene of the crime did not tally with her description of events; a clump of her hair was discovered between Harvey's fingertips; and most telling of all, a bloodstain inside one of her boots provided a genetic match with Harvey's DNA.

Police had retrieved parts of a knife from the crime scene, but had not recovered the weapon itself; the stain in Andrews' boot was about the right size for the blade. They deduced that she had attacked him in a fit of rage, hidden the knife in her boot, and disposed of it later.

Andrews was charged with murder and accused of having attempted suicide shortly after Harvey's death because she was guilt-ridden. Andrews maintained her innocence in court, but the jury found her guilty. She received a life sentence.

inconsistencies discovered later during the investigation and postmortem examination.

The forensic pathologist determines the cause and manner of death to be assigned to the death certificate. This legal conclusion by the forensic pathologist is derived from an analysis of the entire medicolegal inquiry, including the death scene investigation and autopsy and toxicological findings. The information generated may determine whether an individual is charged with a crime, is sued in the civil court for negligence, or receives insurance benefits.

Completion of the autopsy and all the associated paperwork, including the death certificate, is not the end of the forensic pathologist's involvement in a criminal case. A few months or even years later, that case will go to trial. The forensic pathologist, considered an expert witness, will be required to testify to the facts of the case, such as the cause and manner of death. In addition, they may be asked to give an opinion on a variety of hypothetical questions that are pertinent to the case.

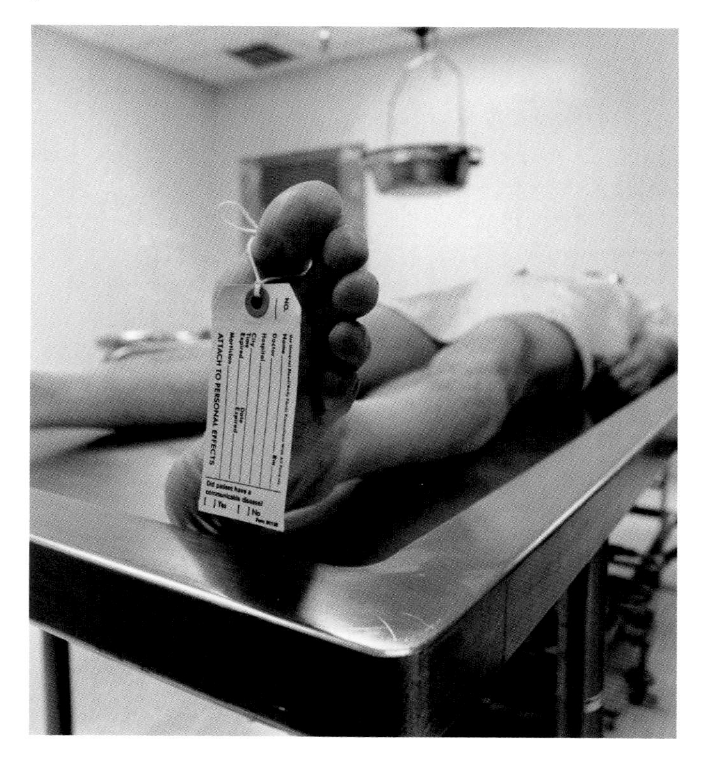

A cadaver on a table awaiting an autopsy where a forensic pathologist submits a sizeable mass of tissue and other samples to the lab for further tests. These samples may include the entire stomach contents, and parts of the liver, kidneys, brain, and a urine sample.

A Day in the Life: Criminal Intelligence Analyst

Nicola Aranyi, Principal Analyst, Lincolnshire Police, U.K.

Being a criminal intelligence analyst is all about having the ability to look at a problem from a wide angle and being able to put the pieces of the puzzle together to make recommendations and inferences that will aid crime prevention or further an investigation.

On a typical day, I begin by looking at the crime picture from the previous 24 hours, using a computer database that logs crime reports from a variety of sources, including the public and police officers. Often I am given set priorities for the types of crime that need attention, such as drug abuse, burglary, or vehicle crime, and I work on my own initiative within those guidelines, assessing if any particular crime trends or hot spots are emerging, and mapping them with specialized computer software.

If any problem becomes apparent, burglary in a specific area, for example, I analyze the modus operandi (MO), study the intelligence picture—who is out of prison, who lives in the area, what information has been received from informants—and set out to provide a total picture for senior officers in the department.

I had noticed, for example, an increase in burglaries in a particular part of town. The MO was the same each time: The offender broke in through a back door using a garden fork or shovel and stole electronic equipment. The burglar often prepared himself food in the kitchen, and clearly was not nervous. I had been mapping the crime picture and monitoring all the intelligence, which centered on a group of teenagers who were suspected of committing up to 200 local burglaries. I ascertained which crimes had produced trace evidence and advised the fingerprint criminalists to run checks.

I presented the information to detectives investigating the crimes, who used it to plan a surveillance operation. To assist this, I provided charts that showed sequences of events, criminal networks, and lifestyle patterns, such as known associates of the suspects, their residences, the cars they drove, and who handled the property. The team was able to carry this work forward and make arrests.

Analysts really come into their own on major cases, such as murder and rape, since they can quickly establish a sequence of events and point to where further investigation is needed. An analyst is brought in at the beginning of a case to collate the movements of the victim on charts and maps; analyze information from witness statements, closed-circuit TV, and other intelligence sources; and reconstruct what took place. Next, an analyst looks for possible offenders, using the same resources.

Once these have been charted, any gaps in intelligence become apparent, indicating where further investigation should be focused. As the case progresses, the analyst is fed a stream of intelligence updates from the investigators, adding detail to the overall picture and assisting in the eventual apprehension of the offender.

03 The Forensic Team

Every crime scene is unique. Soon after the detectives and death investigators arrive, they will assess the scene. Depending upon the apparent type of death, they determine whether specialists need to be called in and if more sophisticated equipment is required to process the crime scene effectively.

The basic team of a death investigator and a forensic photographer may be adequate for a number of situations (a natural death, for example). However, if the scene contains blood spatter or a weapon or if, because of their years of experience, the death investigators and detectives suspect that something looks out of place, criminalists will be called in.

Criminalists process and analyze evidence from the body (such as bullet wounds or residue), evidence from the surrounding scene (trace evidence such as hairs, fibers, blood, fingerprints, tool marks, shoeprints, and DNA), and evidence remote from the scene (such as telephone calls and documents). They collaborate closely with the other investigators of the crime.

The moment a criminal begins to approach a potential victim's residence, office, or other location, he starts leaving forensic evidence. Take the example of a burglary on page 58.

A forensic expert examines the trunk of a car for evidence. Insignificant things such as fibers, hairs, and DNA can be easily overlooked by the criminal as he attempts to cover his trail.

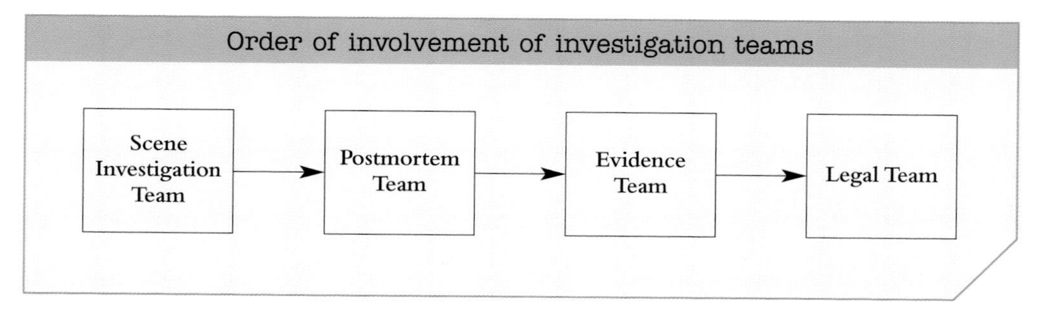

> "During the early hours of the morning, a burglar approached a home that he believed was unoccupied. In order not to be seen by the neighbors, he walked around to the back of the house. Unknowingly, he stepped in some mud, leaving the impression of his shoe. Such impressions can tell a shoeprint expert the type, make, and shoe size, even if the individual walks heel to toe or side to side.
>
> When he arrived at the back door, he removed his sweatshirt, wrapped it around his left hand, and broke a glass panel in the door. When he did this, small fibers from the sweatshirt were left behind on the broken window, and microscopic shards of glass became embedded in the shirt and his skin. He did not even notice the tiny cuts on his index finger, or the small amount of blood he left on the edge of the glass. Once inside, he turned on his flashlight and started searching the home for cash and other valuables. During this entire time, without realizing it, he was shedding hundreds of dried skin cells, hairs, and fibers. As he crept through the downstairs rooms, he was leaving behind clear shoeprints.
>
> The occupants of the house awoke, startling the burglar, who had thought the house was empty. He quickly departed the way he had entered. He might not have left with any of the family's valuables, but he did leave a lot of evidence for the forensic team and detectives to collect and submit for analysis."

Fingerprints Examiner

A latent print criminalist will attend almost every crime scene and will be directed by homicide detectives to the most likely locations to search for prints. Latent prints must be enhanced to make them visible, using a procedure called "developing."

The most common method employed involves dusting a surface with a very fine powder using a camel-hair brush. After dipping the brush into the powder, the criminalist spreads the powder over the surface with light strokes. The powder adheres to the oils deposited by fingertips, outlining

Latent fingerprints can be found with a laser.

The standard method of searching for prints uses dusting powder.

the ridge detail of the prints. The color of the powder depends on the surface being dusted. Black carbon powder is used on light-colored surfaces, whereas prints on glass, silver, and dark surfaces are dusted with a light gray powder. Latent prints can also be detected by subjecting objects to a different light wavelength. Once the prints have been developed, they are photographed, and then transferred or lifted. This is done by applying transparent tape to the dusted area, pressing it down, and peeling it back to remove the powder. Then the tape is placed on a card.

Locating fingerprints at a scene or on an object is only half the battle. They may be of significant relevance or limited value. After being taken to the crime lab, all the latent fingerprints discovered at the scene must be positively identified. The prints may belong to any number of people, including the handyman who came to fix a trash compactor or the victim's boyfriend. The role of the homicide detectives is to ascertain the normal traffic pattern at the scene and obtain fingerprints from all those individuals who could have

Despite the gains made by DNA technology, fingerprinting remains a vital tool in crime investigation. No two individuals' fingerprints are alike.

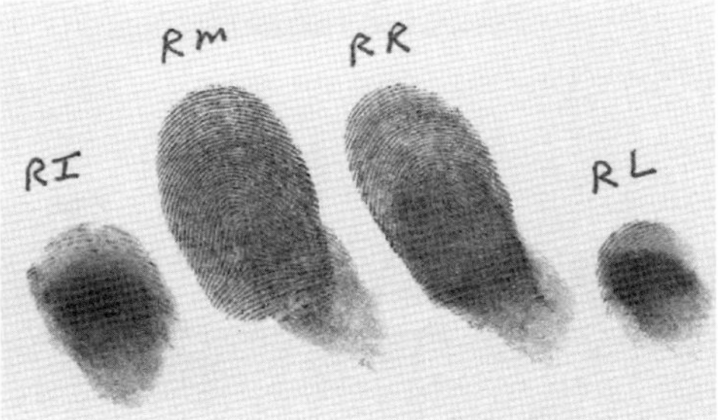

A Day in the Life: Forensic Odontologist

Dr. Michael N. Sobel, Chief Forensic Odontologist, Pennsylvania

I arrive at my private practice orthodontic office early, at 8:15 A.M., to find two messages. As a chief forensic odontologist (forensic dentist), I am on call 24 hours a day, seven days a week. On average, I am given five or six cases a month, although some of them can run for months or even years.

The first message is a request to contact the District Attorney's Office to discuss a bite-mark case that is coming up soon for trial. The second is to call the Coroner's Office regarding a dental identification case where the victim had died in a house fire. I made the calls, agreeing to meet the DA the following week to discuss the bite mark case trial, and to examine the house fire victim later in the day.

By early afternoon, I finish treating my patients and drive to the Coroner's Office, where I examine the burned remains of a white male. My goal is to legally identify this victim from the sets of dental records provided to me. Often, there are delays in obtaining dental records. The name of the victim's dentist must be found from family and friends, and then the actual records must be tracked down from the dentist. In this case, all is in order, and one hour later I have completed my examination, charting and taking digital photographs of the body in question.

I am able to positively match a set of dental records to the victim. Creating a solid identification is relatively easy in this particular case, after I find three root canals, one of which shows an unusually curved root. In addition, there are two crowns, both with porcelain facings. Also, the combination of composite and silver amalgam fillings is consistent with the dental X-rays and charts furnished by the family dentist.

Not all cases are so straightforward. Difficulties arise in putting together shattered body parts that have been scattered, such as in mass disasters, and in locating teeth that have fallen out of the skull during decomposition.

On some occasions, I am called to the crime scene to help search for pertinent dental-related body parts, such as teeth and skull bones. Having been trained in forensic anthropology, I often examine the other craniofacial bones for additional information such as age, race, disease state, and sex. In some bite- and patterned-skin-mark cases, it can be useful to examine the crime scene to better appreciate and relate the evidence gathered.

I file my report with the senior deputy coroner and head home. During the drive, my mind wanders to the bite-mark case. There appear to be some minor inconsistencies between the defendant's set of teeth and the patterned mark configuration. All bite- and patterned-mark cases can be difficult, and I must be constantly careful not to misidentify an imprint as a human bite when, in fact, it is made by something else.

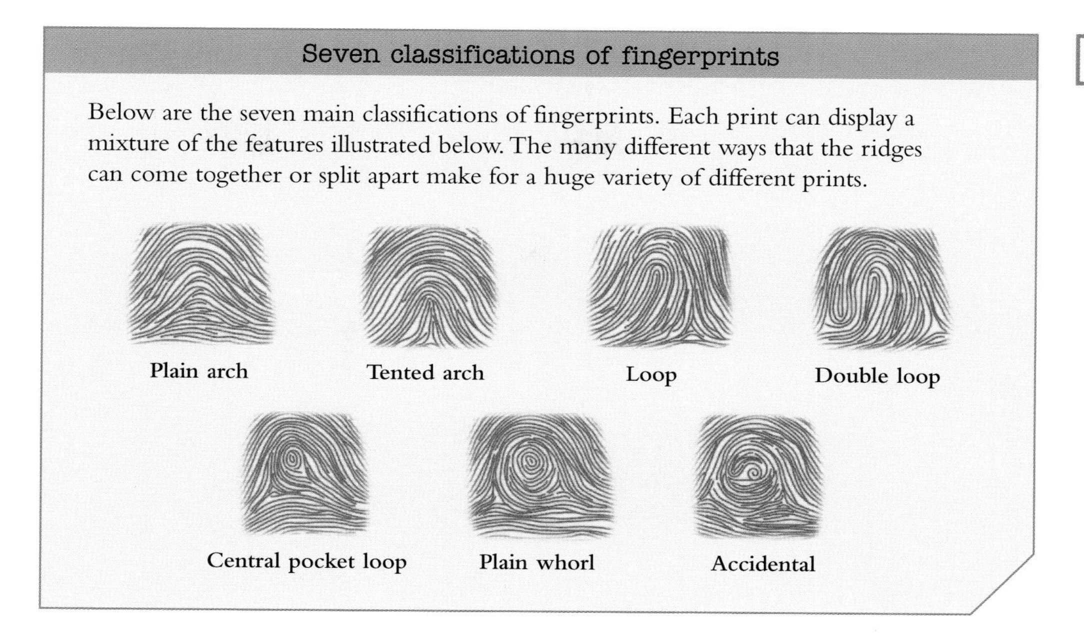

been present. Innocent visitors to the scene will be excluded by comparing their fingerprints to those found during the investigation.

There are seven principal classifications of print pattern: plain arch, tented arch, loop, double loop, central pocket loop, plain whorl, and accidental. In the old days, prints found at the scene were compared to those on file individually by an expert trained in fingerprint identification. This still occurs in some non-modernized police zones. Recent advances in scanning technology, however, have made it possible to digitize fingerprints from the cards and compare them to various databases. In the United States, the FBI has the prints of over 67 million individuals on file; the Canada the Royal Canadian Mounted Police database has 3.5 million sets of prints.

A positive match indicates only that the person who left the prints was at the scene at some time in the past. That said, it is another part of the puzzle, information that will help other members of the investigative team narrow their search. Once a potential suspect has been identified through fingerprints, the detectives need to establish other types of evidence (motive and opportunity) linking that individual to the crime.

DNA Examiner

> "The police report stated that the victim, a young female, had been found beaten and left to die in her apartment. There were signs that a significant struggle had taken place. The only information the police could supply was that the victim had been seen talking with her boyfriend earlier in the day outside her office. During the postmortem examination, a specially trained autopsy technician cut the fingernails from the victim and placed them in evidence containers. These were sent to the DNA criminalist."

Deoxyribonucleic acid (DNA) is one of the most powerful tools in the field of forensic investigation. Essentially, DNA is a large molecule made up of four nucleotides (adenine, guanine, cytosine, and thymine). These nucleotides are chemically linked and form the

Case file:

Richard Ramirez

In Los Angeles between mid-1984 and mid-1985, a series of violent attacks were carried out on people in their own homes in the middle of the night. Men were shot dead and women raped. Satanic pentagrams were left at the crime scenes, and in one case on a victim's thigh. Police knew only that they were looking for a tall, thin man, Hispanic in appearance with a strong body odor.

In August 1985, a victim of the attacker was able to write down the license number of the attacker's car as he drove off. The police discovered the car in a parking lot—it had been stolen at the time of the previous attack—and kept it under surveillance, but no one returned

double-strand helix of DNA. Every human cell with a nucleus contains DNA, and specific parts of the molecule are unique to the individual, making it possible to link DNA found at the crime scene with that of the suspect.

> "Peering through a microscope, the DNA analyst meticulously removed minute blood crusts from the undersurface of the victim's fingernails, using sterile forceps. (Care must be taken not to contaminate the evidence with foreign DNA, since this could jeopardize the analysis and significantly erode its evidentiary value.) The expert's task was to isolate enough DNA from the few crusts found under the nails to produce a genetic profile, sometimes referred to as a DNA fingerprint, which then could be used to identify the source of the blood, possibly the killer."

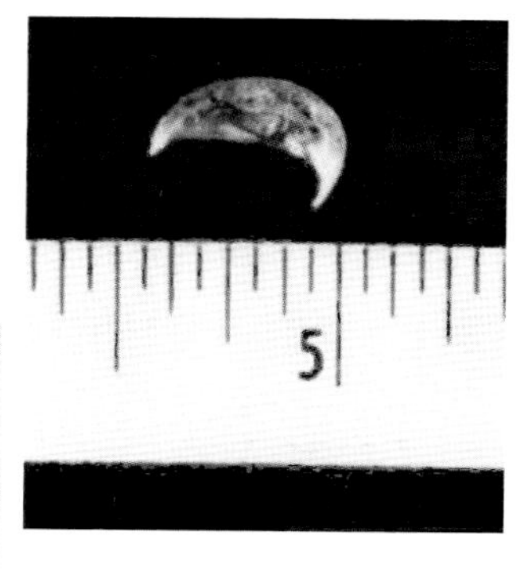

The fingernail from a homicide victim with forensic evidence in the form of skin and blood from the aggressor. This evidence can be used for DNA analysis.

to it. When the car was examined by forensic investigators, they recovered a suspicious fingerprint.

Earlier in the year, the Los Angeles Police Department's fingerprint records had been partially computerized, which simplified the search for the suspect print. The database provided a match with Richard Ramirez, whose fingerprints had been taken some years before following a minor traffic violation. An immediate search was made for him, and his photo was released to the media.

At the time, Ramirez was in Arizona and unaware that the police were hunting for him. When he returned to L.A., he went into a liquor store and was recognized by other customers, who had seen his photo in the newspapers. They chased him down the street and he ran straight into a passing police officer.

Ramirez denied having anything to do with the crimes, but police found the murder weapon in the home of one of his friends, and his sister was in possession of jewelry stolen from the victims. Police attributed 16 murders and 24 vicious assaults to Ramirez. The trial that ensued was one of the longest in U.S. history. Close to 1,600 witnesses were interviewed.

After deliberating for four days, the jury finally came back with a verdict of guilty. Ramirez was sentenced to death in November 1989.

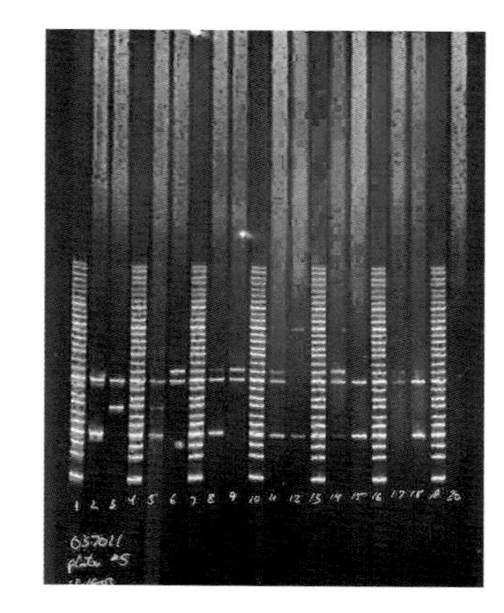

DNA "fingerprints" of several individuals can be examined concurrently to determine if a suspect can be linked to a crime via DNA evidence.

Before genetic testing can be performed on a forensic sample, the DNA must be isolated from any non-biological contaminants as well as from the proteinaceous material that makes up the cellular structure and its components. Since DNA is found in the nucleus of a cell, and red blood cells have no nuclei, the white cells, called leukocytes, serve as the storehouse for the nuclear material in a blood sample. Other sources of DNA include saliva, forcibly plucked hairs, epithelial cells (which line the entire surface of the body and most hollow structures within it), muscle, bone, teeth, and semen.

The extraction or isolation process consists of treating the leukocytes with chemicals that gently break down the cellular and nuclear walls without causing damage to the DNA inside. The DNA is then further purified and concentrated using special molecular filters. This sample is called the DNA extract.

It is imperative that the DNA analyst determines not only how much DNA is isolated, but also how much of it is human. Forensic samples, like fingernail scrapings, often contain bacterial and fungal contaminants, which also contain DNA. Also, an estimate of the amount of human DNA must be made before continuing with the analysis. This technique, called slot blotting, compares the amount of DNA isolated from the unknown sample to standard amounts of human DNA utilizing human molecular probes.

Blood samples are collected from the suspect (in this instance, the boyfriend), and from the victim of the assault, which are used as reference standards during the analysis. Suspects often voluntarily submit blood samples when it is explained to them that they have nothing to fear if they did not leave DNA at the crime scene. DNA can clear a suspect as well as assist in his conviction. The blood samples are spotted onto clean cotton swatches, air dried, and carefully packaged and documented. DNA from small snippets of these blood patches is isolated employing the same chemical processes used on the forensic unknown samples.

DNA analysis: the process

Many forensic samples of DNA are minuscule in size, and much of each sample may be degraded through exposure to heat, humidity, or other harsh environmental factors, leaving little DNA to be tested. Consequently, before analysis can begin, small, well-defined segments of the DNA are replicated millions of times to provide a sufficient quantity to work with. This is done with a process known as Polymerase Chain Reaction (PCR), which mimics the body's ability to copy its own DNA when producing new cells. Each copy of the DNA, called an amplicon, is identical to the original segment and contains all the necessary information to provide a virtually unique genetic profile.

Even though greater than 99 percent of human DNA is the same for all people, approximately 1 percent is unique to the individual (with the exception of identical twins), and this uniqueness is of interest to the DNA analyst, who can exploit the genetic differences to obtain important information about the source of the forensic sample.

The analysis consists of identifying and categorizing the unique features found in the genetic code of the DNA isolated from the forensic sample. These appear as differences in the lengths of the segments of the amplified DNA (PCR products), which vary between individuals.

The amplicons are separated by placing them on a special sieving gel, which allows small PCR products to move quickly through it, while large PCR products travel more slowly. Fluorescent tags attached to the PCR products during the replicating process allow the movement to be recorded with the aid of a laser and a fluorescence detector. By monitoring the exact length of time it takes for the PCR products to pass through the gel, the DNA analyst can compare the results to known standards and assign a genetic profile to the sample.

A bullet casing discovered at crime scene and marked off for examination by forensics experts.

What the genetic profile does

The genetic profile identifies the uniqueness of the sample and distinguishes it from other genetic profiles, much like how a barcode is used to identify a particular product in a store. By comparing the profile of the forensic sample to the profile obtained from the reference bloods, the forensic scientist can provide police and the courts with an estimate of how likely it would be to randomly select a person with the identical genetic profile.

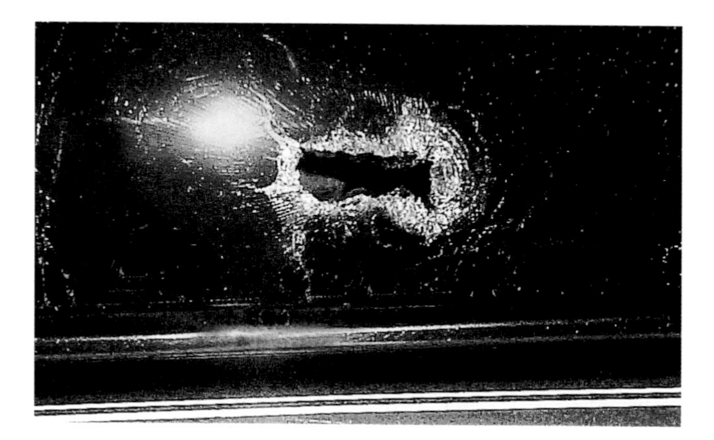

Several bullet holes are shot into the passenger window of a vehicle during a drive-by shooting. Ballistics experts use photographs such as these to examine the angle, distance of the aggressor from the vehicle, and type of firearm used.

In the previous example, the DNA profile obtained from the forensic sample exactly matched the profile obtained from the blood of the boyfriend. It is estimated that the likelihood of two people having the same genetic profile is one in a quadrillion. Put another way, consider a person wandering through a storage facility where there are a million boxes, each containing a billion marbles, one of which is blue while the remainder are white. That person would need to reach blindly into one of the boxes and remove the one blue marble—a very unlikely event.

In the United States, the genetic profile obtained from the forensic sample would also be entered into the National DNA Data Bank, referred to as CODIS (Combined DNA Indexing System). This database contains genetic profiles obtained from all forensic unknown samples as well as from persons convicted of crimes such as homicide, rape, aggravated assault, and other sexual offenses. The profile would be compared with the profiles in CODIS; if a match were obtained, all interested police agencies would be notified so that they could share the results of their investigations. In Canada a national database includes DNA information from young offenders, adults convicted of serious crimes and unidentified samples from crime scenes.

Firearms Examiner

When there is evidence that a gun has been used at a crime scene, the investigators will call in a firearms examiner. This criminalist concentrates on four features of the scene: bullets, shell casings, signs of gunshot powder residue, and the weapon itself. A gunshot wound can be inflicted by rifled firearms (revolvers, pistols, rifles, and machine guns) and non-rifled firearms (shotguns). In a rifled firearm, the inside of the barrel features spiral grooves, which cause the bullet to spin, stabilizing its flight. A non-rifled firearm has a smooth barrel. Handguns, rifles, and shotguns all produce very different and identifiable wound patterns.

If gunshot wounds are apparent on the body, no attempt is made to recover any bullets at the scene. They will be retrieved by the forensic pathologist during the autopsy examination. The pathologist will identify the wounds of the entrance and exit, and the path and direction of the bullet or bullets. These determinations are critical in establishing the manner of death. The key questions in this type of death are:

1. How is the wound size or pattern related to the range and direction of fire and the type of bullet?
2. Can the range of the shooting be estimated from the characteristics of the gunshot wound?
3. Can the relative positions of the victim and the source of the gunfire be determined from the pattern and path of the wound?
4. When several wounds are present, which was inflicted first?

A small-caliber firearm with blood patterns that indicate the direction of fire.

ABOVE A forensic firearm examiner using a bullet comparison microscope.

The job of the firearms examiner includes determining the locations of the shooter and the victim and the path of each bullet. The last is accomplished by using strings or laser lights.

Information in the pathologist's autopsy report assists the examiner in determining the correct positions of the shooter and victim. This reconstruction method also helps locate bullets that may be embedded in walls, ceilings, and the firewalls of vehicles.

Drive-by shootings typically leave a large number of shell casings at the scene. In these cases, the street will be blocked off by patrol officers while homicide detectives carefully search for the casings. When a casing is located, its position is marked and photographed, and the distance between the casing and the victim measured. When all the casings have been located, they are collected and transported to the ballistics crime lab.

Bullet holes in glass windows are commonly encountered. By examining them carefully, it is possible to determine the direction from which the shots were fired and the sequence in which the holes were made. As a bullet passes through glass, it makes a small hole in the outer surface, but leaves a larger hole in the inner face. This enables the direction of fire to be ascertained. Where there are several bullet holes, the sequence of the shots can be assessed by

BELOW Comparing the lands and grooves of two bullets under a comparison microscope.

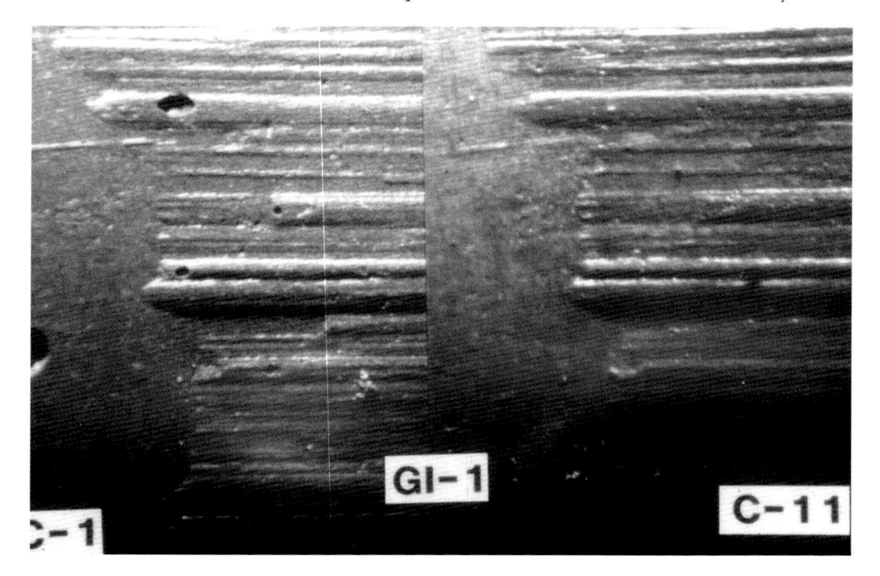

examining the intersection points of the fracture lines in the glass.

The role of the firearms expert does not end at the scene. By analyzing the bullets and casings, the examiner will be able to determine the number and type of weapons used. If a weapon was located at the scene, it will be test fired into a tank of water to obtain a defect-free bullet for comparison purposes. The test bullet and bullets recovered from the scene and/or body are placed in a comparison microscope, which allows two objects to be examined at the same time. Bullets fired from the same weapon display an identical pattern of grooves, caused by their passage through the barrel. If a match is found between a scene bullet and the test bullet, the weapon can be identified positively as having been used in the crime.

IBIS: the Integrated Ballistics Identification System. Originally used only in the United States, ballistics information is now high on the agenda of many governments around the world, and data is now shared between many cooperating countries.

Shell casings also bear two important markings: those imprinted by the manufacturer and those caused by the firing process. After the test bullet(s) and cartridge cases have been compared to bullets and cartridge cases in the open case file, their unique features will be entered into the Integrated Ballistics Identification System (IBIS). IBIS is a database that allows them to be compared with those recovered from other crime scenes, making it possible to identify weapons that have been used in more than one crime. Initially used only in the United States, information about bullets and cartridges is now shared internationally since the 9/11 attacks.

Testing for gunshot residue

When an individual fires a handgun, particles from the primer and powder of the ammunition may be deposited on his or her hands. For this reason, the death investigators seal the hands of the victim with paper bags to protect and preserve any gunshot residue (GSR). At the morgue, prior to the internal examination, a specially trained autopsy technician conducts a test for GSR. This involves wiping the palms and backs of the victim's hands with cotton swabs treated with a few drops of a 5 percent nitric acid solution. The interiors of any shell casings recovered from the scene will be treated in the same manner. Each swab is sealed in its own vial and the vials sent to the crime lab. There the

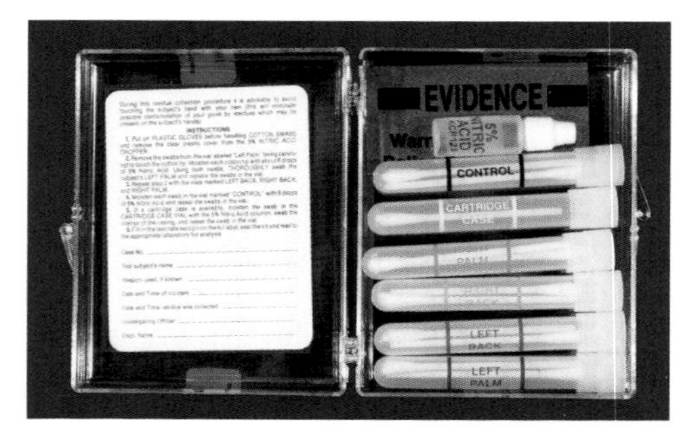

An atomic absorption kit used to test for gunshot residue.

swabs are tested for the presence of chemical elements found in the primers of most cartridges. If these are discovered, it is highly likely that the person had discharged or handled a gun shortly before death.

If the victim was wearing clothing during the shooting, the criminalist will examine it carefully for firearms-related evidence, such as bullet holes, and signs of soot and gunpowder residue. In addition to a visual inspection, the criminalist will use infrared photography and microscopic examination. The distance between the weapon and the victim can also be determined by comparing the powder particle pattern on the clothing to test patterns produced in the laboratory.

Hair/Fiber Examiner

Hair and fibers are found at many crime scenes because they are easily shed from humans, animals, and clothing. This trace evidence can be collected by special vacuum cleaners designed to sweep carpets, furniture, and automobile floors and seats. These devices have filters that trap fragile physical

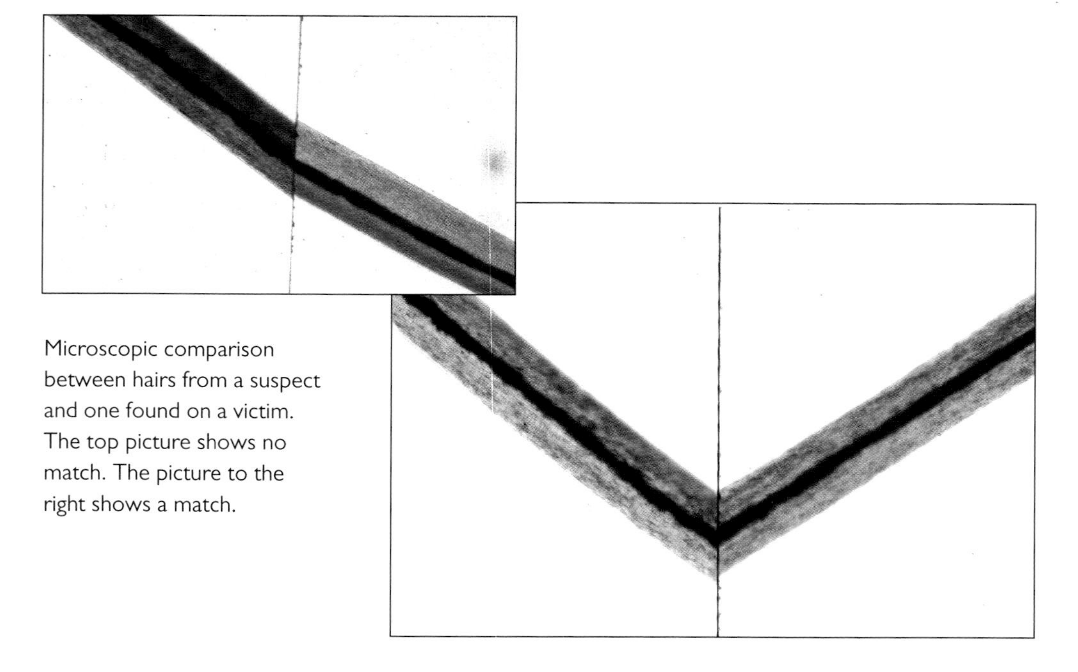

Microscopic comparison between hairs from a suspect and one found on a victim. The top picture shows no match. The picture to the right shows a match.

A Day in the Life: Forensic Toxicologist

Dr. Frederick W. Fochtman, Chief Toxicologist, Pennsylvania

As a forensic toxicologist, each day I find myself looking at a variety of blood, body fluids, and organ tissues submitted from recent deaths of undetermined origin. Opening the containers cautiously, I remove portions to test. In some cases, it can be immediately apparent that the person has been dead for some time.

I prepare the samples for analysis with a variety of instruments, such as immunoassay analyzers, gas and liquid chromatographs, gas chromatography/mass spectrometers, gamma counters, and head-space analyzers. After the instruments produce all their data, I have to interpret the information and come up with answers regarding the cause of death.

I may be asked by police officers to evaluate specimens and cases for driving under the influence of drugs and/or alcohol. Again, I have to evaluate and interpret the data to provide a report that can be used by the police to prosecute the offender. There are times when I find myself in the courtroom testifying for either the prosecution or the defense. Even though the analytical results cannot be disputed, there may be some interpretation that could aid one side or the other.

I may also investigate workplace incidents following positive drug testing. I am asked to review each case, interpret the results, and give an opinion as to whether the presence of drugs contributed to or caused an accident. One case involved a worker who had suffered a minor fender bender. Company policy was that any accident had to be followed by a drug screen, which proved positive for morphine. I studied the case and noted that the driver claimed to have eaten almost all of a box of crackers before reporting for work. The crackers in question had poppy seeds baked into them, and poppies are the source of morphine, so I arranged an experiment where I asked volunteers to eat the same crackers and I collected their urine afterward. All of my volunteers tested positive for morphine in the urine. Fortunately, new testing criteria means that poppy seed ingestion will rarely, if ever, produce a morphine-positive result in the future.

A new prescription drug has been approved recently by the United States Food and Drug Administration and is being given frequently to patients with chronic pain. There have been reports of misuse of the drug and possible overdoses. Today, specimens have been submitted from a case that is suspected to be a lethal overdose of this drug. To proceed, I will have to search the literature for information about analytical methods and case reports. I will need to obtain a sample of the drug to use as a standard, and prepare various reagents to properly validate the assay. Upon analyzing the specimens, I will have to interpret the results. This can be difficult with a new drug when not much is known about toxic and lethal blood concentrations.

evidence. Another method of collecting hair and fibers is to press wide strips of adhesive tape onto the area of interest. When these are removed, they lift the evidence with them. The strips are placed between two layers of plastic to protect the evidence while it is transported to the lab.

Hair and fiber evidence is frequently collected from hit-and-run incidents. The examiner will closely inspect the clothing and lower extremities of the victim. Samples of fibers are retrieved, classified, and stored until the vehicle is recovered. Then the criminalist will conduct a detailed examination of the grille, bumper, and headlight areas for hair and fibers. If any are found, they are compared to the trace evidence gathered from the victim. The criminalist uses a comparison microscope (similar to those used in ballistic comparisons) to make a side-by-side comparison.

One of the tasks of the hair/fiber examiner is to differentiate between human and animal hair. This is done by conducting a microscopic examination of the hair, inspecting its three basic parts: cuticle (scales), cortex, and medulla. As a rule, matching hair found at the scene to a suspect generally requires 12 full-length hair from different areas of the suspect's head and pubic region.

Fibers are common trace evidence, primarily because of the ease with which they transfer from one surface to another. They can be natural (wool or cotton, for example) or synthetic (rayon, nylon, or the like). Natural fibers are easily recognized under the microscope by their medullas and cuticular cells; man-made fibers are identified by their dyeing patterns, diameter, number of strands per thread, weave construction, twist pattern, and number of twists per inch.

A blood spatter pattern. From this evidence experts can reveal that the drops fell from a height of 70 ft. (21.3 m) and at a 90-degree angle with the impact surface.

Bloodstain Examiner

The laws of physics dictate what happens when blood leaves the body. A drop of blood is spheroid or ball shaped, not the common teardrop shape often depicted. The typical blood drop is approximately 0.05 ml or 50 microliters and attains a terminal velocity of about 25.1 ft. (6.6 m) per second.

The analysis of a bloodstain pattern includes an evaluation of the shape of the spots, their size, and the surface they hit. The last is most important: the same amount of blood hitting different surfaces will cause very different spatter patterns. The harder and less porous the surface, the less the blood will spatter. In contrast, blood hitting

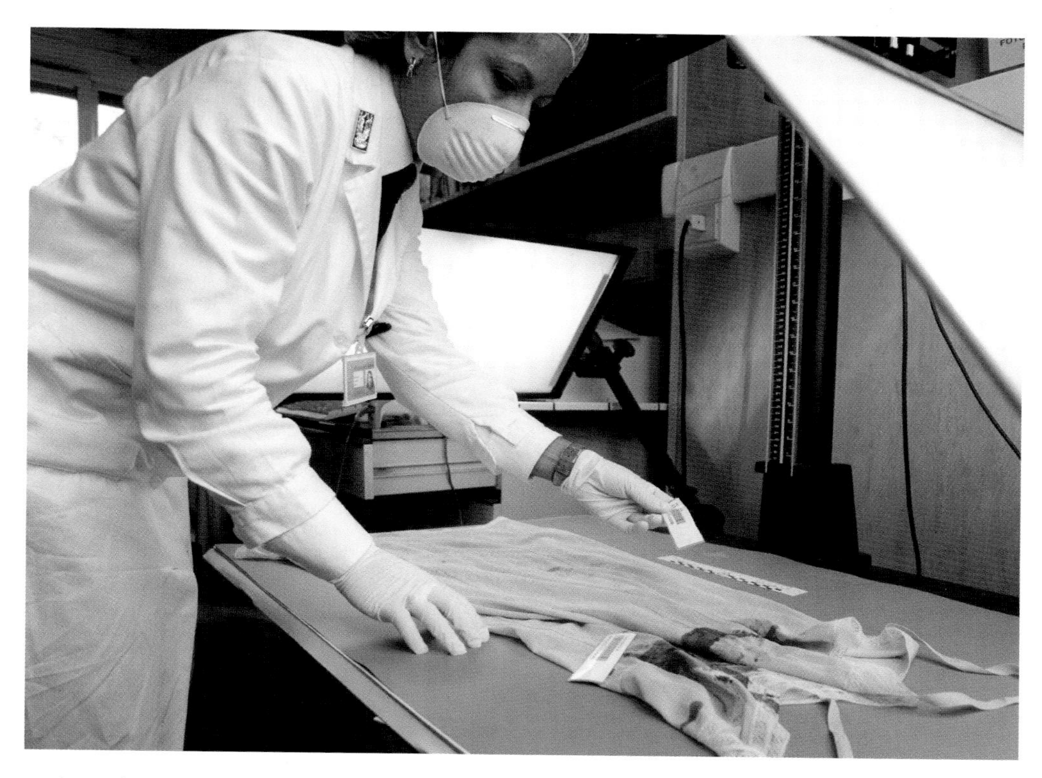

an irregular, porous surface will result in greater spatter. Therefore, bloodstain evidence is meaningless unless the surface texture of the point of impact is known.

The shape of the blood drop after striking the surface will be either round or elliptical; the direction in which it was traveling can be determined from the shape. The origin of the bloodstain pattern is determined by drawing lines through the long axes of the individual blood spots. The point where these lines converge gives a two-dimensional location of the point of origin. However, it does not provide any help in ascertaining the height of origin. The diameter of the bloodstain is of little or no value: the size of the drops is dependent on the type of surface they impact, not the distance they fall. The shape of the blood drops gives an idea of the angle at which they hit the surface. The angle of distortion from a geometric circle is indirectly proportional to the impact angle. There are three types of impact: low-, medium-, and high-velocity. Each produces unique, characteristic stain patterns.

Finding blood at a crime scene has been useful since the discovery of blood groups at the beginning of the twentieth century. This kind of analysis allows detectives to narrow their suspect pool considerably. Now DNA analysis is performed on blood samples instead.

> "Police were called to a bloody crime scene. A woman had been assaulted while she was getting her mail. She had been stabbed in the throat with a knife, dragged inside her home, and placed in a chair. The motive appeared to be robbery. When the police found her, they noticed a trail of dried blood running down her left arm, which had formed a pool surrounded by smaller droplets on the tiled floor."

The example above given by a bloodstain examiner is a typical example of a low-velocity impact wound. It is caused by blood dripping onto itself; liquid-to-liquid impact results in a large pool surrounded by small droplets.

The clotting and drying times of blood can be used to determine the time an offense was committed. Blood usually clots between three and six minutes after it has been shed.

Case file:
Wayne Williams

During the two years between mid-1979 and mid-1981, Atlanta was plagued by a brutal serial killer who was strangling young black men and boys. In their search for the murderer, police had little to go on apart from some forensic evidence in the form of a particular type of carpet fiber found on the bodies. When the news of this evidence leaked out, however, the strangler began stripping his victims and dumping them in rivers so that any fibers would be washed away.

The police immediately began patrolling the rivers, and in the early hours of May 22, 1981, officers by the Chattahoochee River heard a splash and saw a car stopped on the Jackson

> "An argument between a newly married couple resulted in a fatal outcome when the husband struck his wife's head with a baseball bat. According to the statement given by the 19-year-old husband, his 18-year-old wife had accused him of having an affair, which started a bitter verbal argument. He claimed that he had been just swinging the bat to let off steam and accidentally had hit his wife once. Photographs were taken and an analysis of the bloodstain patterns was conducted."

A medium-velocity impact produced by a beating creates bloodstains with diameters of 1 millimeter or larger. In this case, the number of impacts, not the weapon, was in question. To answer this query, the criminalist needs to look for evidence of cast-off bloodstain patterns. During a beating with an instrument, blood does not come to the surface immediately. After blood covers the area, it will produce spatter, and blood will adhere to the beating instrument.

Parkway bridge. They questioned the driver, 23-year-old Wayne Williams, but had no concrete reason to detain him, so he was allowed to leave. Two days later, the naked body of 27-year-old Nathaniel Cater was recovered from the river.

Williams was arrested, and when the carpets in his home were subjected to forensic examination, fibers matching those taken from the victims were found. Additional samples of the same fiber were retrieved from the trunk of his car. When police traced the manufacturer, they got their big break. They learned that less than 16,500 square yards of the carpet had been made in the same color, and the odds of it being in another house in Atlanta were almost 7,800 to one. Furthermore, other fibers discovered on the bodies matched those of the carpets in Williams' car, and fewer than one in 3,800 cars in the city would have had the same carpets. It was estimated that the chances of anyone else in the Atlanta area having the same combination of carpets in their home and their car as Williams would have been 30 million to one.

Williams was tried for the murder of Cater and one other victim. He was found guilty on both counts and received two consecutive life sentences.

As this is raised and swung back by the attacker, it produces multiple cast-off patterns along the path of travel. The minimum number of blows can be established by counting the individual cast-off blood trails. One additional blow should be added to the total because the first blow rarely produces a cast-off pattern.

Document Examiner

Not all crimes are solved by looking for changes to the body during a postmortem examination or gathering trace evidence at the scene. Sometimes, a careful study of documents, especially legal documents, can reveal changes that point the way to a likely suspect.

For example, a greedy son may change the will of his mentally impaired father to benefit from a greater inheritance. This is done by creating a new will and forging the individual's signature. Normally, such cases come to the attention of law enforcement agencies after the reading of the will, when other members of the family may become suspicious because of their knowledge of the kind of relationship that existed between the father and his son.

In these cases a document examiner is called in to assist the detectives. This criminalist's role is to make a critical examination of the document in question and compare it with other documents to:

1. Establish that it is genuine.
2. Expose forgery in the forms of alterations, additions, or deletions.
3. Identify individuals through documents showing the authorship of handwriting.
4. Provide testimony when required.

As a rule, where written documents are concerned, the longer the document, the harder it is to accomplish undetected forgery. Usually, as in the case of a will, the signature is the most common type of attempted forgery. While the typed section of the will may also be in question, the forged signature is easier to detect.

Analyzing a Document

The document in question is placed in a thin, transparent plastic folder, which does not interfere with the visual

Forging a signature

The basic methods of forging a signature are:

1. Freehand drawing by copying an adjacent example.
2. Freehand drawing from memory.
3. Tracing from a genuine copy using a transparency, superimposition, or transfer of a genuine signature.

examination. It is marked as an exhibit, and a reference number is placed away from any written area. In addition, the document is photographed.

The examiner can use various methods to analyze a document, starting with a direct visual examination. The use of magnifying lenses, microscopes, and other optical aids, including radiation, infrared, and ultraviolet devices, will follow. The paper and ink are also chemically analyzed.

Examples of genuine Salvador Dalí signatures have been used to authenticate unknown pieces of the painter's work.

In the case of a will, the signature is compared to known genuine signatures of the deceased. These are obtained from uncontested documents such as canceled checks, letters, and employment applications. If the detectives suspect an individual, that person will be asked to submit a sample of handwriting, either voluntarily or through a court order. The examiner will compare the rhythm and flow in the execution of the signature. Forged signatures, when compared to the genuine item, show inhibited, unsure, and irregular rhythm of writing, defects in line spacing, hesitation, pen pauses, differences and inconsistencies in pen pressure, slow speed with thick beginning and ending of strokes, retouching, and presence of guidelines.

In court, an examiner will give his opinion as to whether or not the signature in question is genuine. He may qualify his conclusion by describing it as beyond a reasonable doubt, possible or probable. In the example of the son forging his father's will, the document examiner noted that the signature on the will looked exactly like the signature on a business letter discovered by the detectives at the son's home.

While we all learn to write essentially in the same manner, we develop our own unique styles during adolescence. Once established, an individual's writing style remains relatively stable throughout life. However, a signature is never precisely the same each time. Therefore, if two signatures are exactly identical, they may indicate a forgery. In the case of the forged will, the son traced his father's signature from the business letter. This was corroborated by high-powered magnification, which showed evidence of superimposition of the signature from the letter onto the will.

Aside from analyzing handwriting, document examiners identify typewriting, ink, and paper. They also attempt to establish the data, source, history, and sequence of preparation, alteration or additions to a particular document.

Other Types of Criminalists

There are many other types of criminalists, each of whom has a specific expertise in processing evidence that can

Case file:

Stephen Bradley

In June 1960, Bazil and Frieda Thorne, who lived in Sydney, Australia, won a large sum of money in the state lottery. Five weeks later, their eight-year-old son, Graeme, was kidnapped while on his way home from school. Shortly after, a man with a heavy foreign accent phoned them to demand a $52,500 ransom. After a second call, nothing more was heard.

The police mounted an immediate search for the boy, but at first all they could find were his schoolbag, cap, coat, and books. Then Graeme's body was discovered on August 16, 10 miles from his home and wrapped in a rug. He had been suffocated and bludgeoned to death.

When forensic scientists examined the rug and boy's clothing, they discovered minute traces of pink cement mortar, dog hair, and a variety of plant matter. Moreover, mold found on Graeme's shoes and socks suggested that he had been dead for about six weeks; in other words, he had been murdered shortly after being abducted.

Subsequent analysis showed that the dog hair was from a Pekinese, while among the plant debris were the seeds of a rare cypress tree that did not grow anywhere near the body's location. The police began an immediate search for houses with pink mortar joints and the rare cypress growing nearby, even going to the extent of issuing a public appeal for any information about

contribute to the successful investigation of a crime. These include anthropologists, whose specialty is the identification of skeletal remains; odontologists, who use their knowledge of dental matters to identify victims; entomologists, who can help in determining time of death from insect activity; and botanists, who can provide important clues when plant matter is associated with a death scene. Even an accountant may be required on a forensic team if a victim's financial affairs could have some bearing on the reason for the crime.

A forensic scientist works on a DNA sample in a genetic fingerprinting laboratory at Leeds University in Britain.

The criminalist's job does not end at the crime scene or in the laboratory. These specialists also play key roles during the pretrial preparations and the trial itself. As experts in their specific fields, they are often called to testify in court to explain to the jury the procedures they followed, and also to provide an interpretation of their findings and conclusions.

such combinations. Eventually, they discovered one in the suburb of Clontarf. When they interviewed the tenants of the house, they learned that its previous occupant had been a Hungarian who used the name Stephen Bradley. He had spoken with a pronounced accent and had kept a pet Pekinese dog. Moreover, he had owned a blue 1955 Ford Customline that matched the description of a car seen near the kidnap site on the day that the boy had disappeared.

When the police carried out a painstaking search of the house, they unearthed a photograph of Bradley's family picnicking on the same rug that had been found with the boy's body, together with a tassel that had fallen from it. Their inquiries revealed that Bradley had sold his car on the day that he had disappeared, but they managed to trace it to a nearby dealer. When the car was subjected to forensic examination, grains of pink mortar were found in the trunk. Further investigation led the police to his dog, which had been left in the care of a local veterinarian—its hairs matched those found on the boy's clothing. By then, Bradley was on a ship heading for England, but the police had it intercepted at Colombo, Sri Lanka. He was arrested and returned to Australia for trial, being found guilty of the boy's murder. For this, he received a life sentence.

04 The Autopsy

Once a coroner's office takes jurisdiction of a body, it is obliged to conduct a death investigation or autopsy (postmortem) examination. This can be either an external examination, a partial, or a complete autopsy.

The forensic pathologist carries out the examination and decides on the type of procedure necessary to provide the appropriate answers. In making that decision, he or she will review the death scene investigative report, police reports, medical records, and any other relevant information (a suicide note, for example).

Five questions of the forensic pathologist

The goal of the examination is to answer:

1. Who is the deceased?
2. What is the cause of death?
3. What is the manner of death?
4. Is there any health risk to the community?
5. In cases of potential litigation, has all pertinent evidence been located, documented, and preserved?

Types of Autopsy

There are three basic types of autopsy examination: external, partial, and complete. The first involves a detailed head-to-toe

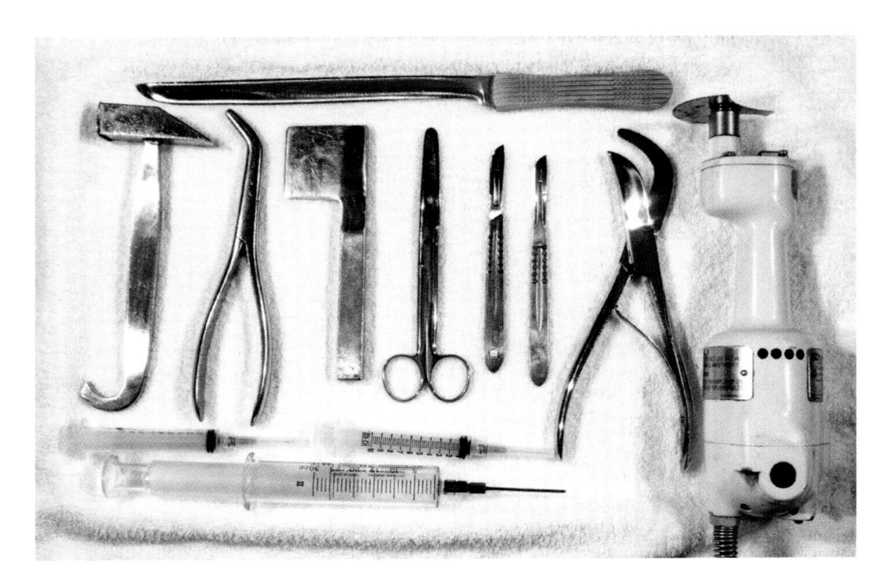

Autopsy instruments: the hammer and a hook, the dura-stripper and the wedge are used to remove the skull; the scissors and scalpels are used to make the "Y" incision for the removal of organs; the rib cutter removes ribs and chest plate; the saw is used to cut through the skull to the brain; the knife is used to section the organs; and the needles are used for collection of body fluids.

external inspection of the body, during which the forensic pathologist will note all scars, deformities, trauma, and postmortem changes to the body. The external examination may be used where the cause and manner of death are obvious, either by the appearance of the body (in cases of hanging or self-inflicted wounds to the head, for example), or because the deceased is elderly or infirm and there is significant medical documentation of a natural cause of death, such as heart disease.

On occasion, an external examination will be conducted if the family of the deceased person is strongly opposed to a complete autopsy for religious or ethical reasons. Even so, body fluids (blood, urine, and eye fluid) will be recovered for toxicological analysis.

A partial autopsy involves the external inspection plus a limited detailed examination of one area of the body. For example, a study may be made of the brain to determine the presence of a subdural hematoma (blood clot) or Alzheimer's disease; the neck organs of a fire victim to obtain evidence of soot in the airway; the chest to demonstrate bronchopneumonia or a myocardial infarction (heart attack); and the abdomen to check the liver for cirrhosis.

Which type of autopsy?

Upon arrival at the autopsy suite each morning, the forensic pathologist will review the day's caseload in the chief autopsy

A Day in the Life: Forensic Epidemiologist

Dr. Steven A. Koehler, Chief Forensic Epidemiologist, Pennsylvania

Unlike other members of the investigative team, I never get the 3.00 A.M. call to attend a death scene. Later in the morning, however, copies of the investigative reports of the deaths that have occurred overnight will be waiting for me. On a typical day, there may be a homicide, a fire victim, and two drug overdoses. All need to be entered into the coroner's computer database, which has been designed specifically for death investigation.

The coroner's office collects a vast amount of information about the victim in each case: age, sex, race, internal organ weights, and the results of the toxicological analyses. In instances of homicide, details of the suspected killer are also assembled. A death investigation report prepared by the deputy coroners contains a full description of the circumstances of the death. It is my job to take this large amount of intelligence and enter it into the database.

Next, I put the information to use by providing it in a meaningful form acceptable to the Food and Drug Administration (FDA), law enforcement and public health agencies, and various medical/legal professionals.

It is essential to summarize and present the data in a way that is readily understandable by the public. For example, with homicide cases, I don't simply give the total number of homicides in a year; I break the figure down by age, sex, race, and location to highlight the most vulnerable section of the population. I also dig deeper into the motives for these killings and am able to show, for example, that over 50 percent of homicides in our county arise from domestic disputes.

I will plot the location of any homicide on a map and provide this information to the local law enforcement agency.

I monitor the level of drug overdose deaths and, if there is a significant increase from one year to the next, I will prepare a report to the FDA to make them aware of the tendency. I also conduct trend analyses of the data to locate those areas with emerging drug problems, then provide this information to the Graduate School of Public Health at the University of Pittsburgh so that they can target their intervention and prevention programs.

My busy day may be interrupted by a phone call from a local news reporter, asking perhaps for the latest information on the number of deaths from fires. The reporter may also want historical data on the number of fatal house fires in the past five years and their causes. After searching the database, I will send a report.

When reviewing the deaths examined by my office, I am always looking for cases that are unusual, or that offer additional information and knowledge regarding a particular type of death.

Another responsibility is to collaborate with the forensic pathologists and other experts in preparing manuscripts for publication in peer-reviewed journals. I also write articles, present abstracts, and speak at national conferences.

technician's office, where all the information for each case is kept. This will include a copy of the death investigation report, medical records if the individual died while in the hospital, the autopsy fact sheet/diagnosis, the autopsy protocol, and a working copy of the death certificate. The autopsy fact sheet will be filled out with things such as weight, height, eye color, clothing, and scars. After studying this information, the pathologist will decide upon the type of autopsy required.

The following is typical of the kind of death investigation report reviewed by the forensic pathologist:

> "The 45-year-old white male was in his residence at 1:45 P.M. when he was discovered by his landlord to be unresponsive. He was lying face up on the living room couch. The landlord then called emergency services. Eight minutes later, paramedics and police arrived.
>
> The examination determined that the man was already dead, and no resuscitation was attempted. The victim was pronounced at 1:56 P.M. The police officer at the scene notified the coroner's office about the death. At 2:30 P.M., two DIs arrived, each dressed in blue pants, white shirt with a badge, and a black jacket with the word CORONER in reflective white lettering. They noted the victim was lying on the living room couch dressed only in shorts and a T-shirt. The victim's girlfriend became concerned when she could not reach her boyfriend by phone. She last spoke with him the night before around 9:30 P.M. The girlfriend stated that he had seemed to be depressed for the last few days.
>
> The scene appeared to be in order. There was no drug paraphernalia found in the residence. No suicide note was found. Positive identification of the deceased was made by the landlord. The core liver temperature was 85°F (29°C) at 2:45 P.M., and the room temperature was 70°F (21°C). No trauma or foul play was noted at the scene. The body was placed in a body bag, put on a stretcher, and set in place within the coroner's van. The apartment was sealed and the landlord was instructed not to enter the residence until the next of kin had been contacted."

After reviewing this particular report, the pathologist determined that a complete postmortem examination was necessary.

The Complete Postmortem Examination

Once the decision has been made on the type of autopsy required, the morgue comes to life. The lights of the examination room are turned on, a phone call is made to the photography department to summon a photographer, and the autopsy technicians change from their street clothes to scrubs. A technician enters the cooler and checks the names on the body bags to locate the body needed for examination. The cooler is a large refrigerator that maintains the body at 40°F (4°C); some can hold up to 75 bodies at a time. Before being placed in the cooler, each body is weighed and an ID photograph taken of the face.

The body is retrieved from the cooler, removed from the body bag, and placed on a stainless steel examination table. The first task is to photograph the body in the condition in which it was found. The pathologist describes and records the

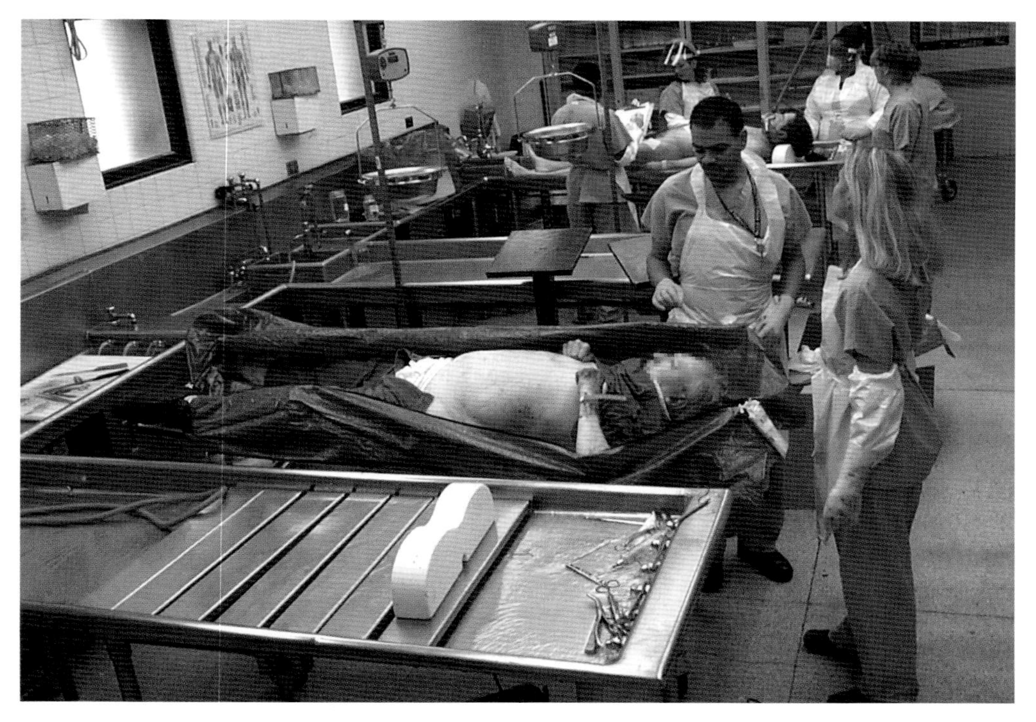

Bodies at the morgue being examined by a forensic pathologist and autopsy technicians.

type, color, and condition of the clothing, if any, and notes the presence of any jewelry. Then the clothing is removed and the body photographed again.

The clothing and jewelry are taken by the technicians and placed on large white sheets of paper, which are folded carefully to prevent any trace evidence from being lost. These personal effects are passed to the forensic science laboratory for examination.

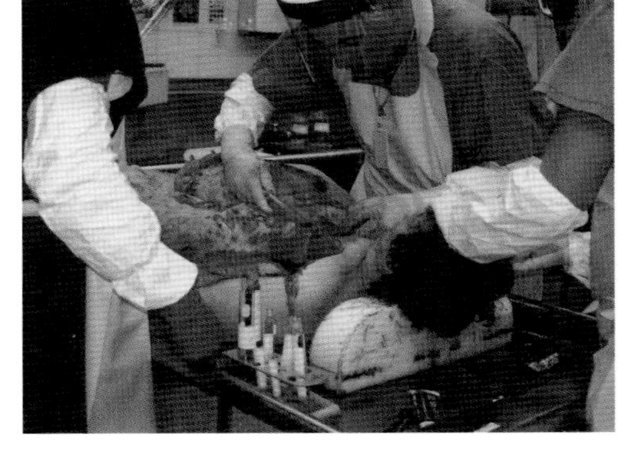

The pathologist examines the body in great detail from head to toe, looking for any evidence of trauma (recent and remote), scars (old and new), deformities, tattoos, and signs of natural disease. While not as unique as fingerprints, tattoos can be used to aid positive identification. The length of the body is measured, and the extremities are flexed to determine the degree of rigor mortis. Other information such as hair and eye color, condition of the teeth, and the presence of petechial hemorrhages (tiny red or purple spots) in the conjunctivae (transparent membranes) of the eyes is recorded.

TOP During the internal examination, blood is removed from the heart and sent to a toxicologist for examination.

ABOVE Bile is removed from the gallbladder for toxicological analysis.

The "Y" incision

Once the autopsy technicians have set up the instrument tray, the body can be opened up using the classic "Y" incision. This provides access to the chest plate, the liver, and the large and small intestines. The chest plate is removed by cutting through the 12 ribs on each side of the sternum with a rib cutter, or bone saw, therefore exposing the heart and the lungs.

Prior to the removal of any organs, the thoracic abdominal cavities are inspected and body fluids are withdrawn for toxicological analysis. A needle is inserted into the heart or its great vessels (aorta or pulmonary artery) to collect blood. Another needle is used to draw off urine from the bladder located in the pelvis. The liver, the largest organ within the abdominal cavity (weighing on average 60 oz./1,500 g) is pulled up to expose the small green-colored gallbladder on its posterior surface to collect bile. The last body fluid obtained for toxicological analysis is eye fluid (vitreous humor).

The internal organs—heart, lungs, liver, spleen, and pancreas—are removed and weighed individually. After the stomach has been taken from the body, its contents are examined for pills, and signs of ingested substances and food. The large intestine is over 5 ft. (1.5 m) long, and the small bowel is approximately 23 ft. (7 m) long. These are taken out, opened, and examined. The pelvic organs are withdrawn: the bladder, prostate, and testes in males; the bladder and internal reproductive organs in females. These are followed by the kidneys and the adrenals, which are located in an area called the retroperitoneal space. Finally, the aorta is extracted, and any soft tissues and fluids are cleared out so that the pathologist can inspect the cavities to look for any abnormalities, which will be photographed if present.

Next, the scalp is peeled back, the skull opened, and the brain removed for examination. If hemorrhages or other signs of trauma are encountered, the photographer is called on to document them. Finally, the neck organs are examined, first in place and sometimes with layered dissection of the muscles in cases of suspected strangulation. Then they are removed.

Case file:
Dr. Harold Shipman

The body of Kathleen Grundy, an 81-year-old ex-mayor from Hyde, England, was found on a sofa in her home, on June 24, 1998.

Her friends immediately called Dr. Harold Shipman, who had visited the house a few hours earlier and was the last person to see her alive (he said that he had been collecting samples for a survey on aging). Shipman pronounced her dead and the news was conveyed to Grundy's daughter, Angela Woodruff. The doctor told the daughter that an autopsy was unnecessary because he had seen Ms. Grundy shortly before her death.

However, following her mother's burial, Ms. Woodruff received a phone call from solicitors who claimed to have a copy of her mother's will that left £386,000 ($550,000)

Dissection of the organs

The pathologist carefully inspects and dissects each organ taken from the body, and has them photographed, if necessary. Small portions of each organ are taken for examination under a microscope after being "fixed" in formalin, which preserves the tissue. Small parts of the preserved tissues are used to prepare histology slides; the remaining tissues are kept in formalin storage for several years.

The heart is the first internal organ to undergo examination. Its weight varies with the weight of the individual, and there is a simple method for calculating the correct weight. All that is necessary is to take the body's weight in pounds, multiply it by two and call it grams. That is the approximate weight of the heart. For example, in a 150-lb. (68 kg) individual, the heart should weigh around 300 g.

The heart receives its nutrition and oxygen via the blood from the coronary arteries that course across its surface. If the muscular structure of the heart is damaged, it loses its ability to

to Dr. Shipman. Ms. Woodruff, believing it to be a fake, went to her local police.

An autopsy was ordered which, in turn, required an exhumation order from the coroner. The body was removed to the laboratory, where normal autopsy procedures were followed; blood, tissue, and hair samples taken from Ms. Grundy's body were sent to different labs for analysis.

At the same time, police raided the doctor's home and offices, and found an old Brother manual portable typewriter that Dr. Shipman told them he had sometimes lent to Ms. Grundy. Later, forensic scientists confirmed it was the machine used to type the counterfeit will and other fraudulent documents.

Meanwhile, the forensic toxicologist, filed her report, showing that there was a lethal level of morphine in the dead woman's body and also that Ms. Grundy's death would have occurred within three hours of receiving the overdose. Dr. Shipman's use of the drug was a serious miscalculation, since morphine is one of the poisons that can remain in body tissue for centuries.

Dr. Shipman later claimed that Ms. Grundy was a morphine addict, but as Dr. Shipman's past patients were exhumed and more autopsies were carried out, toxicologists found high levels of morphine in each of the bodies.

The doctor was given consecutive life sentences for the murder of 15 people. It is thought that his victims numbered more than 400. In January 2004 Dr. Shipman died in jail, apparently by suicide.

Autopsy

During an autopsy, a Y-incision is made from the body's shoulders down the middle. After folding back the skin and opening the rib cage, the pathologist concentrates on the following organs for examination.

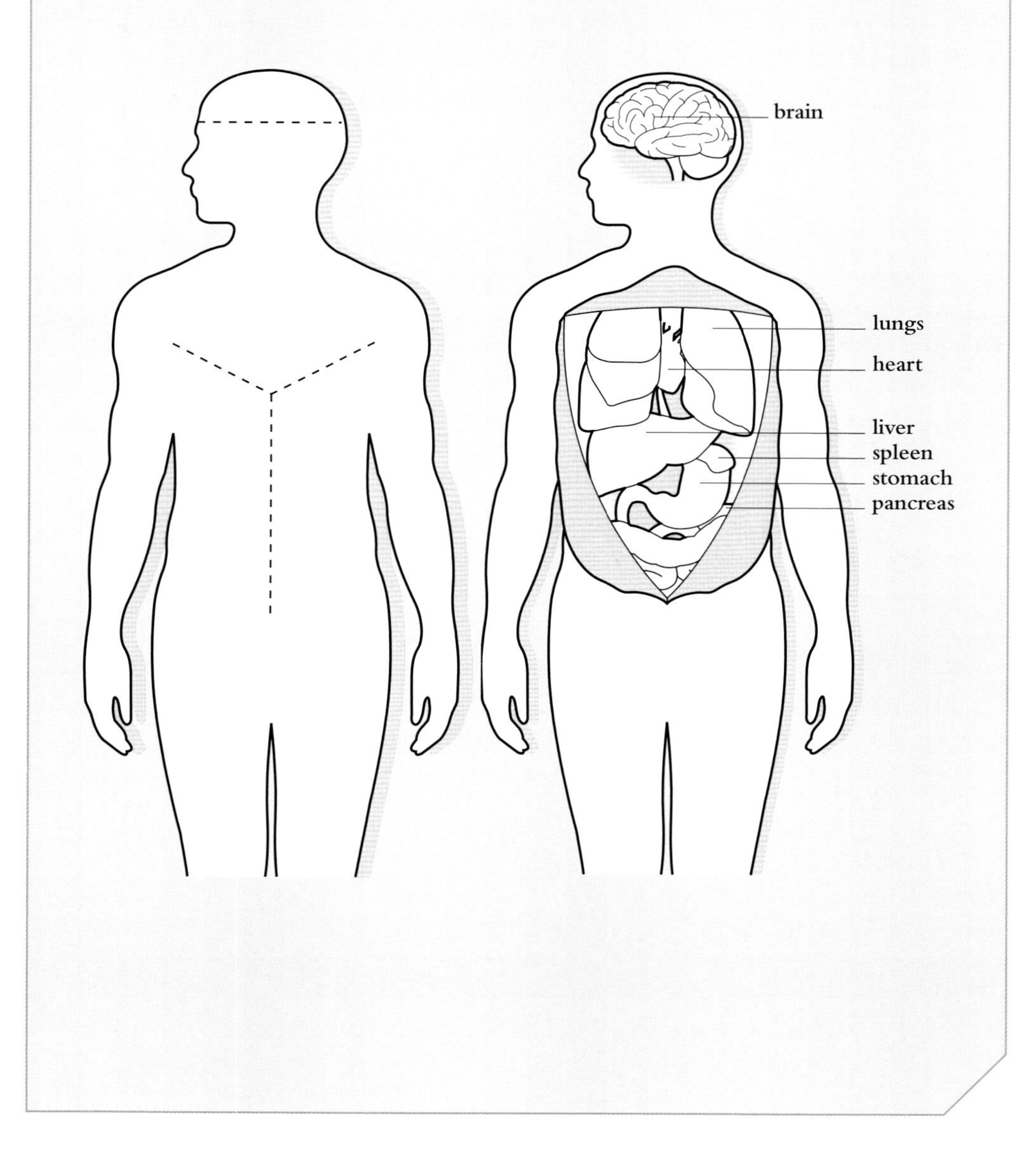

pump blood to the other vital organs of the body. Damage or death occurs in the muscle when one or more of the three main coronary arteries are blocked. This is known as coronary artery disease. It is caused by small plaques blocking the flow of blood, which in turn causes a part of the heart to die (hence the pain). If the blockage causes muscle damage, which leads to diminished cardiac function, the result is decreased blood flow to the brain, necrosis (death) of the heart muscle (myocardial infarction), and ultimately death of the individual.

There are many other abnormalities of the heart that can be found, including signs of infection (myocarditis), changes due to hypertension (thickening of the left ventricle of the heart), and valvular heart disease (such as aortic stenosis or mitral valve prolapse). By far the most common cause of natural death seen in a coroner's office is atherosclerotic cardiovascular disease (narrowing of the coronary arteries).

The next organs examined are the lungs. These are located on each side of the heart; the right lung consists of three lobes and weighs up to 450 grams, while the left lung has two lobes and weighs up to 350 grams. The function of the lungs is to deliver fresh oxygen to the body, and remove carbon dioxide and other gaseous waste products. The inhaled oxygen passes through the mouth into the trachea (windpipe), then to either the left or right mainstem bronchi of the lungs, and on to the smaller bronchial branches. These split into even narrower branches and ultimately terminate in 300 million minute air sacs called alveoli. At the alveolar level, the blood-gas exchange takes place. Lung diseases that are frequently encountered include pneumonia, emphysema, and cancer.

The next most common cause of death is related to disorders of the central nervous system. This includes the brain, a pale-colored organ consisting of two cerebral hemispheres, the cerebellum, and the brain stem, all made up of nerves and weighing about 1,500 grams. Hemorrhage on the brain, typically due to rupture of a blood vessel, and inside the brain, frequently the result of hypertension (abnormally high blood pressure) or stroke, is commonly discovered.

If the heart, lungs, and central nervous system are normal, the focus of the autopsy switches to the minor organs, such as the liver and pancreas.

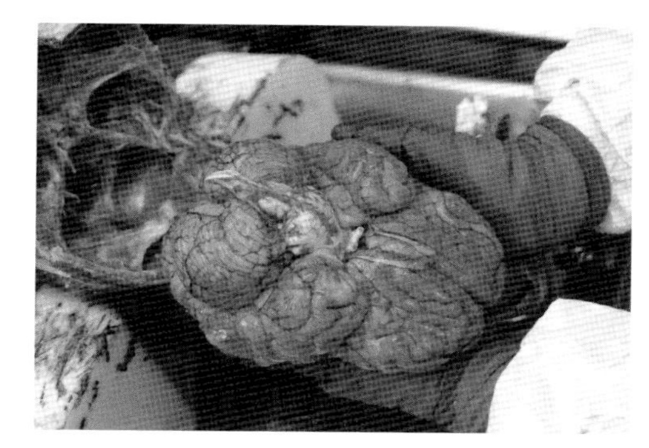

The brain is first examined by the forensic pathologist for signs of trauma, hemorrhage, and evidence of disease such as Alzheimer's. After this inspection the autopsy technician cuts the spinal cord, and the brain is removed and weighed.

In a significant number of cases, the examination of organs will fail to reveal a cause of death. In this situation, the forensic pathologist will consult the forensic toxicologist for the findings of the body fluid analyses.

After all the internal organs have been removed and examined, they are put in a red biohazard bag, which is placed in the chest cavity. The sternum is replaced and the body is sewn up.

Forensic Toxicology

Toxicology is defined as the study of poisons and their effects on the human body. A forensic toxicologist is interested in the chemical analyses of materials, both physical and biological. Usually, forensic toxicologists are employed by coroners' offices, health departments, and police laboratories. Often, they are involved in a consulting capacity in criminal and civil cases.

In addition to the body fluids collected during the autopsy, such as blood, bile, urine, and eye fluid, in certain cases, the forensic toxicologist may be sent stomach contents and sections of lung, liver, and brain for analysis. Also, they analyze materials collected at the scene, such as bottles containing medications and syringes.

The main tasks of a forensic toxicologist are to identify and quantify toxic substances such as chemicals, drugs, and poisons in biological material. They do this using a variety of analytical methods and specialized equipment.

An analysis of blood for drugs or chemicals will result in one of the following:

- No drugs found.
- Drugs found, with the following levels noted:
  Therapeutic blood level: The concentration of drugs or chemicals in the blood is within the therapeutically effective dosage.
  Toxic blood level: The concentration of drugs or chemicals in the blood that is associated with toxic symptoms in humans.
  Lethal blood level: The concentration of drugs or chemicals present in the blood that has been reported to cause death.

A researcher places samples in an auto-sampler during a toxicology study.

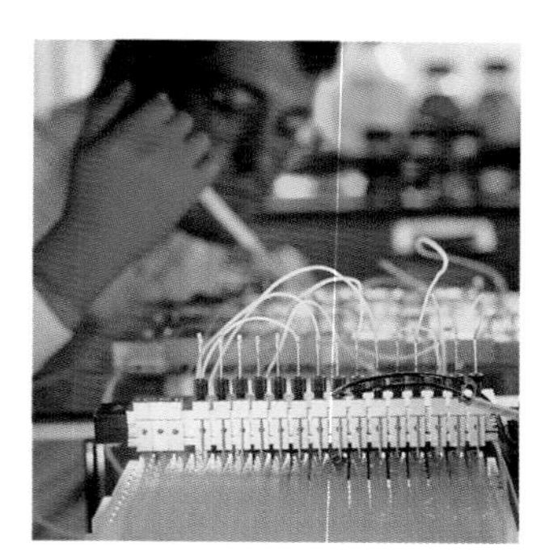

If a lethal drug or chemical level is found in the blood, it does not automatically mean that the death is drug related.

A Day in the Life: Forensic Pathologist

Dr. Shaun Ladham, Forensic Pathologist, Pennsylvania

One of the fascinating aspects of my job is that I never know what type of case I will be investigating on any particular day. Listening to the newscast on the way into work sometimes offers a clue. Today, there is a reported homicide, and, as on most occasions, little detailed information is given. When I arrive at the office at 7:45 A.M., it's normally quiet. The deputy coroners are settling into their 7:00 A.M.–3.00 A.M. shift, and the receptionist's desk is still unmanned. But today, the place is buzzing. I am met by two homicide detectives in the lobby. I ask if they are here for the case reported on the news. They nod and say it's bad. I quickly change into scrubs and walk to the autopsy suite with them. I pick up the paperwork for the case, which includes the death investigation report, and read the following:

"The victim is a 6-month-old white female, who was discovered in her bedroom by police after daycare workers became concerned when the little girl was not dropped off at 6:00 A.M. this morning and there was no answer at the residence. At 6:15, police arrived at the residence, entered, and located the girl DOA. A quick examination of the scene revealed trauma to the head region. A search of the room failed to locate any object out of place or with blood on it. The core body temperature was 97°F (36°C) and the ambient air temperature was 78°F (26°C)."

I examine the photographs taken by the death investigators at the scene. The room is a typical little girl's room, apart from the blood spatter on the bed and the ceiling. I call the forensic science lab and am told that the criminalists are still processing the crime scene. The detectives inform me that the girl lives with her father at this residence on weekdays and at her mother's home during the weekend. The parents are married, but are involved in a tumultuous divorce. No records of past domestic problems are on file with either the police or childcare agencies.

After reviewing all the information, I proceed with the postmortem examination. The girl is removed from the cooler and placed on the autopsy table. Even with my years of experience and having performed thousands of autopsies, I am not always prepared for such brutal and senseless deaths as this. As I examine the little girl, I determine that she was killed by multiple blows to the head with a hammerlike object. I instruct the detectives to return to the scene to search for the weapon. My rage at this senseless killing has to be tempered if I am to conduct a thorough, objective, and complete examination. If not, I could miss critical evidence that might create a problem for the district attorney during subsequent criminal proceedings. The worst part is that, in several months, I will have to relive this experience during my testimony at the homicide trial.

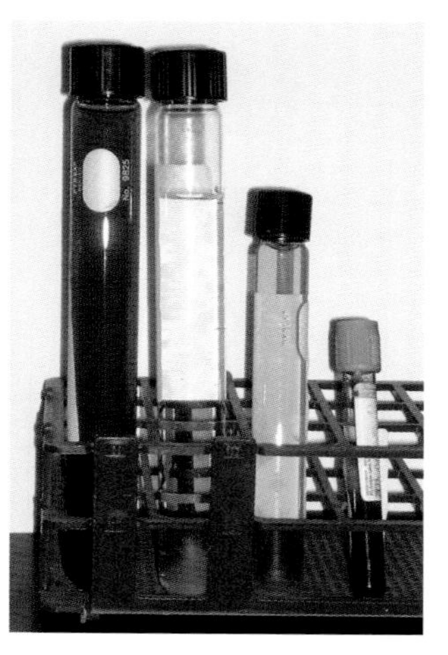

Cardiac blood, urine, bile, and femoral blood collected from a body and stored for toxicological examination.

For example, a known narcotic addict may be found shot to death. Although the analysis of the blood would show a lethal morphine level, in this case, the cause of death would be listed as a gunshot wound to the head. In contrast, take the example of a woman who attempts suicide with barbiturates; she is discovered alive and taken to the hospital for treatment. She dies 15 hours after admission. Postmortem blood levels indicate an amobarbital level of 0.75 mg, which is below the lethal level. However, reference to the medical history will determine that the cause of death was barbiturate poisoning. In this instance, the hospital would be contacted to discover if any blood taken upon admission had been saved. That level would be more relevant.

One of the most frequent tests requested by police agencies seeking scientific evidence of an offender having driven under the influence of alcohol and by pathologists dealing with fatal motor vehicle accidents is the ethyl alcohol level. Alcohol is readily absorbed into the bloodstream from the stomach and the small intestine. The rate of absorption depends on factors such as the concentration of alcohol and the presence of food in the stomach. Water slows the absorption rate, whereas carbonated beverages, champagne, and sparkling wine increase it. From the small intestine, the alcohol passes into the blood and then through the liver, where it is metabolized. The major method of assessing the degree of intoxication is to measure the concentration of alcohol in the blood.

The most common type of unnatural death investigated by a coroner's office is the drug overdose. Of the wide number of substances encountered in overdose cases, heroin is at the top of the list.

Collecting Physical Evidence from the Body

During the autopsy, aside from examining the body and organs, the forensic pathologist is responsible for identifying and collecting other types of evidence from the corpse. This includes tire-track marks, bullets, bite marks, signs of rape, and medical implants.

By far the most common items of physical evidence collected are bullets. The trajectory of a bullet inside the body is affected by its angle of entrance, the position of the body, and whether it strikes any bones. Therefore, the normal procedure is to take X-rays of the body to locate any bullets lodged in it. Once a bullet has been recovered, it is photographed, then placed in an evidence envelope labeled with the case number, the name of the deceased, the location from which the bullet was recovered, and a description of the bullet. It is sent to a ballistics expert for analysis.

During the autopsy, medical implants may be identified. The most common are pacemakers, artificial hips, pins, and breast implants. Each of these devices carries unique information; for example, they may bear a serial number and the name of a manufacturer, which can aid in the positive identification of an individual.

Forensic evidence does not always take the form of metallic objects; in rape cases, there may be bite marks and semen to be collected. If bite marks are found on a body, the site is protected until a forensic odontologist can examine them.

Tires can leave imprints on clay, sand, snow, and bodies. These are often seen in hit-and-run deaths, typically as patterned bruise impressions. Although they are not as individual and definitive as an individual's DNA or fingerprints, tires do have unique wear patterns, tread patterns, and other special features. Any impressions found on a body can be compared to patterns in reference libraries kept by the different tire manufacturers and law enforcement agencies.

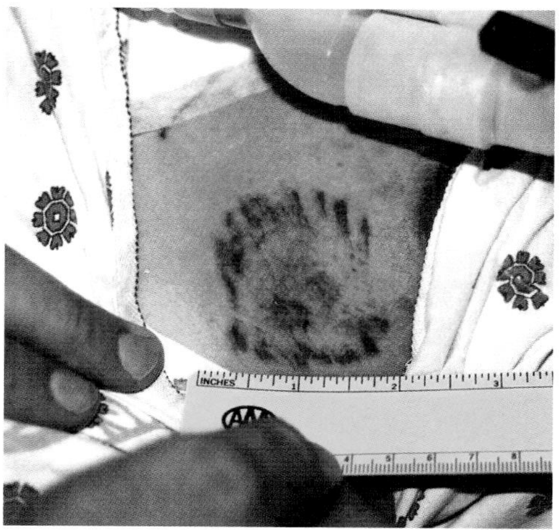

A suspected bite mark on a victim is photographed immediately with a scale before the impression fades.

The Stages of Death

Not all bodies are discovered shortly after death. Many are dumped in remote locations or deliberately concealed under floorboards or in abandoned buildings and, when found, may be in varying stages of decomposition. The degree of decomposition of the body will be affected by its location (inside or outside), whether it is clothed, the ambient temperature, and access of insects, rodents, and other scavengers. While decomposition of the body increases the complexity of the homicide investigation, there are forensic experts who can be called upon to help. These experts are trained in the identification of bodies that are burned beyond recognition or are nothing more then a skeleton. In addition, they can also determine the approximate time of death.

The following hypothetical story illustrates the stages of decomposition and describes the different techniques and specialists employed to determine the identity of the victim and the time of death, from minutes after death to complete skeletonization.

> "A verbal argument between two brothers over their father's will escalated when the younger brother pulled a .38-caliber semiautomatic pistol from the desk drawer. As the clock struck 8:00 P.M., he fired a single shot. The bullet struck the older brother in the head, and he fell to the floor dead."

Death is not a single event, but rather a process. The precise time of death is only known for individuals who are closely monitored, such as a person who dies in the emergency room during CPR. This situation is the exception, not the rule. In the vast majority of deaths occurring during

Case file:

James S. Vlassakis et al.

On May 20, 1999, police officers from a missing persons task force operating out of Adelaide, South Australia, entered a disused former bank in the farming community of Snowtown, north of the city. For a year, they had been investigating a number of disappearances going back to 1993. When they opened the vault, they discovered six black plastic barrels of acid, which held human body parts from eight individuals. They also found handcuffs, knives, ropes, rubber gloves and a machine that could be used to produce electric shocks.

The next day, police raided addresses in northern Adelaide and arrested three men: John Bunting, Mark Haydon, and Robert Wagner. All were charged with murder. Subsequently, two more bodies were found buried at Bunting's home. On June 2, a fourth suspect was arrested, 19-year-old James Vlassakis.

Postmortem examination of the remains indicated that many of the victims had been tortured before they had been killed. Some had gags in their mouths and ropes around their necks; others had been mutilated by having their limbs and feet hacked off; several of the bodies showed evidence of burns. Because of their condition, identification required the combined forensic skills of pathologists, anthropologists, fingerprint experts, DNA analysts, and odontologists. Even though most of the body

a criminal activity or accident, the recorded time of death can never be certain. Criminal cases may hinge on establishing opportunity. Therefore, the estimate of when the death took place will convict some and release others.

If the brother had called emergency services right away, determining the identity of the victim and the time of death would have been relatively easy. An examination of the deceased by a forensic pathologist at 9:00 P.M. would have found a warm body, with all extremities flexible and a core body temperature of 97°F (36°C).

During this early phase of death, visual identification by the next of kin or close friends is possible. In addition, photo identification cards such as drivers' licenses can be used. Most large coroner's offices have a viewing room where family members can see the deceased through a window or by means of closed-circuit TV in order to make a positive ID.

The exact time of death is determined by the forensic pathologist who conducts the autopsy. For the first couple of hours after death, the processes of rigor mortis (body stiffening) and of livor mortis (blood pooling) are unlikely

parts had been submerged in acid, it was still possible to extract sufficient DNA for checking against that of relatives of the missing persons. One breakthrough was that police were able to identify skeletal remains that had been unearthed by a farmer some five years before. In this case, anthropologists were able to compare the bones with an X-ray of the missing individual.

It transpired that all the victims were either related to or known by the accused, and had been killed so that the murderers could fraudulently claim their welfare or disability payments. In some cases, the killers had even impersonated some of their victims when dealing with banks. Before being killed, some of them had also been forced to leave messages on their telephone answering machines to allay the fears of relatives and friends at their sudden disappearance.

Vlassakis admitted to killing four of the victims, including his halfbrother and stepbrother, and was sentenced to life imprisonment in 2001. However, he cooperated with the investigators and testified against Bunting and Wagner at their trials in 2003.

Bunting was charged with 11 counts of murder, Wagner with seven; both received life sentences. Haydon was charged with 10 counts of murder, and is still awaiting trial.

Changes after death

Minutes after death: core temperature rises due to bacterial action
1 hour: body is warm, extremities flexible, core body temperature equal to the normal body temperature of 98.6°F (37°C)
2–6 hours: rigor mortis, livor mortis, set in. If the body is exposed, flies lay eggs in the body
24–36 hours: muscle contraction subsides
3 days +: autolysis and putrefication set in, strong odor, body is swollen, greenish black in color, eyes bulging
6 days +: flies, maggots present

to be observed. Also algor mortis (body cooling) is not a reliable means of assessing time of death, as the temperature of the deceased at the moment of death will be unknown, and body temperature is thought to rise shortly after death.

> "Instead of calling emergency services, the younger brother froze, then decided to move the body to a spare bedroom until he could work out what to do next. As he dragged the body, the skin was warm to the touch, and the arms and legs moved easily. He placed the body face up in the corner, spread a sheet over it, and closed the door."

After death, the body progresses through overlapping stages of decomposition that allow a broad estimate to be made of the time of death.

> "The next morning, the brother tried to move the body outside to a shed. He noticed that it was cool to the touch."

Another method used to establish time of death is core body temperature. Normally, the human body maintains an internal core temperature of 98.6°F (37°C). After death, this internal temperature increases initially, due to bacterial action.

However, after that, the temperature decreases at an average rate of 1.5–2.5°F (0.8–1.4°C) per hour during the first 6–8 hours, then at 1°F (0.6°C) per hour until the body equalizes with the ambient air temperature. Death investigators from the coroner's office normally take a core temperature shortly after arriving at the scene, along with a reading of the ambient air temperature. The surface on which the corpse is lying is also noted; a body on a concrete floor will lose heat faster than one on a sofa.

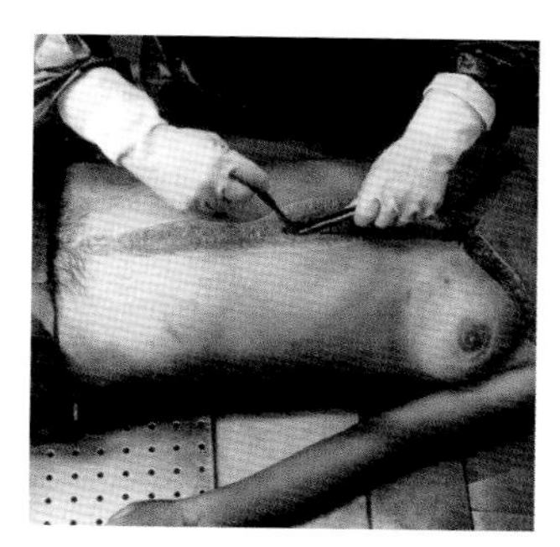

An autopsy is performed on the corpse of a 20-year-old woman. The woman has been dead for two days.

> “During the night, rigor mortis had begun.”

Normally, rigor mortis is first observed 2–6 hours after death, the process being complete between 12–24 hours after death. The muscle contraction (stiffening) subsides approximately 24–36 hours after death. This timetable is very general and can be affected by many variables.

> “With great difficulty, the brother carried the body to the shed and placed it face up on top of some rags. He noticed that when he removed the bloody shirt, there were areas of deep purple discoloration on the deceased's back.”

After death, the circulation stops and blood begins to pool due to gravity. This settling of blood is called lividity or livor mortis, and it causes the skin to develop a pinkish or purple appearance one or two hours after death. For the first 8–24 hours, the livor is not fixed. In other words, pressing on the lividity will cause it to blanch (disappear) for a short time. Also, if the body is repositioned or turned, the discoloration will shift to the new dependent area. In fixed livor, however, the blood will not redistribute even if the body is shifted to a new position. This feature is important in determining if a body has been moved after fixed livor has been established. For example, if an individual dies face up in bed, but 24 hours later is placed face down on the floor to make it appear that he has fallen out of bed, the fixed livor will tell a different story.

> "Several days passed and the brother noticed a smell coming from the shed. As he approached it, the odor became more intense with every step. Opening the shed door, he saw a swollen greenish black body with bulging eyes. The skin was starting to slip off the body. The smell was more than the brother could take and he quickly closed the door as blowflies buzzed around the shed."

Decomposition is the process of autolysis and putrefaction. Autolysis is the breakdown of the complex proteins and carbohydrates into simpler chemical compounds. Putrefaction is the breakdown of tissues by bacteria. The action of the bacteria produces large amounts of foul smelling gas.

At this stage, visual identification can still be carried out, using physical features other than the face. These include such obvious characteristics as missing or extra toes or fingers, missing limbs, surgical scars, and even tattoos. If the body

Case file: Ted Bundy

In a killing career that lasted almost a decade, Ted Bundy is thought to have murdered between 40 and 50 young women. His crimes began in 1969 in California, then he headed to Oregon, Washington, and Utah. He would carry out a number of murders in each area, dump the bodies often hundreds of miles away, and move on before he was detected.

In Salt Lake City in late 1974, he lured Carol DaRonch into his car and attempted to attack her with a crowbar, but she escaped. When Bundy was caught some months later, she was able to identify him and he was sentenced to 15 years in jail. Over the next couple of years,

is beyond visual identification, circumstantial evidence such as clothing or papers found on the victim can also be used.

> “A few more days passed before the brother summoned up the courage to return to the shed. When he opened the door, he could see it was filled with flies, and a strong, nauseating odor of death hung in the air. He took a quick look at the body and saw that thousands of small, white maggots were crawling over it.”

If the body was discovered at this point, the special talents of a forensic entomologist would be employed. Forensic entomologists study insects and other bugs, such as spiders, ticks, and mites, and call flies “the unwilling members of the forensic team.” Their discipline is closely associated with criminology and, consequently, violent crimes. They use their knowledge of insect life cycles to assist in determining the length of time since death—the time commonly termed postmortem interval (PMI)—based upon the examination of

however, he managed to escape twice. Although he was recaptured quickly the first time, he remained free on the second occasion.

On the run, Bundy headed for Florida, but instead of lying low, he began murdering again. In early 1978, he attacked several young women students at Florida State University, leaving two dead and two seriously injured. He was eventually captured in Pensacola, where he had murdered a 12-year-old schoolgirl. Even so, he had covered his tracks well, and was only caught because he had been stopped for drunk driving.

Bundy was sent to trial for the murders of the Florida students, but pleaded not guilty.

During the autopsy of one of his victims, however, a bite mark had been noted on her left buttock. As is normal practice, this was measured and photographed carefully. Bundy's teeth were also photographed and he was compelled to allow a cast to be made of them. When the characteristics of his teeth and jaw were compared to the bite wound, there was no doubt that he had inflicted it.

This proved to be the vital piece of evidence for convincing the jury to return a guilty verdict. Bundy was sentenced to death, but due to numerous appeals he was not actually executed until 1989.

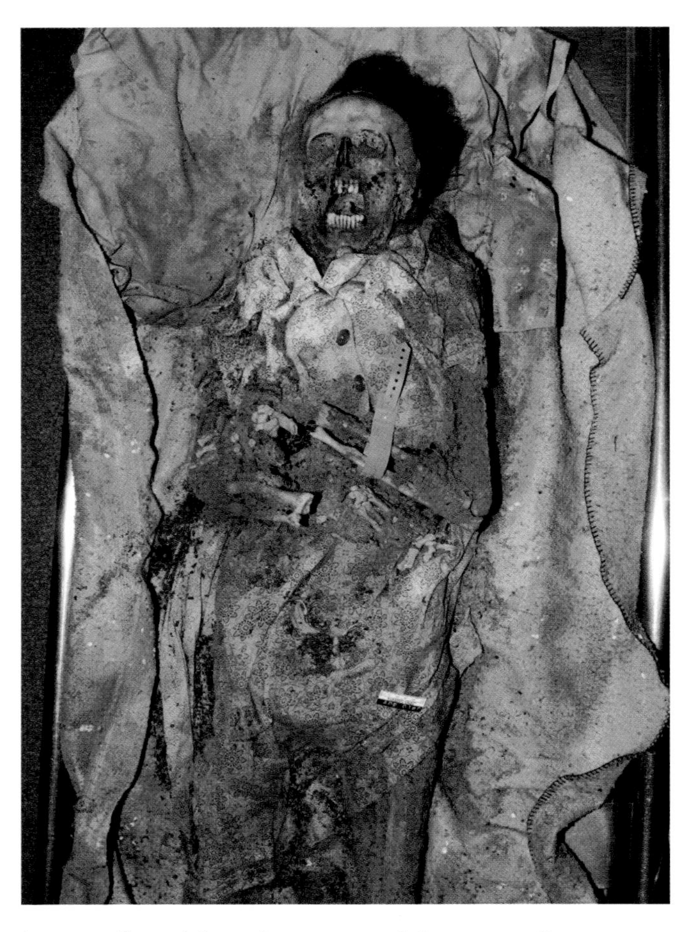

A postmortem examination of a skeletonized body after years buried underground is a challenge for a pathologist because of the level of decay of many parts of the anatomy.

insects collected from the corpse and the surrounding scene. The forensic entomologist uses two main approaches to estimate the PMI:

1. Analysis of the colonization succession patterns of insects
2. Determination of the life stage of the insects collected on, or near, the body

The first method relies on the forensic entomologist's knowledge of the sequence of decomposition and the well-defined order of organisms that infest the body at different stages of that process. It is used when the body has been decomposing for a long period of time. The second method is based on the development time of the fly. Moments after death, the first organisms that usually arrive are flies, most commonly the blowfly (bluebottle fly). They enter the body through the nostrils, ears, mouth, eyes, or traumatized areas, laying eggs that develop into maggots. The maggots are collected from the body and reared to adulthood, when they can be identified microscopically. The estimate of PMI is determined by the known development time of the species found at the scene.

> "Many years later, on his deathbed, the 96-year-old brother confessed to the murder and told his nurse that he had buried the body under the tennis court. The next day, a team of forensic investigators arrived at the court. Among the white-suited investigators were a forensic anthropologist and a forensic odontologist. After a few hours of excavation, they unearthed a section of plywood, and below it a small piece of what appeared to be a blanket and a large number of bones. The bones and surrounding soil were collected and transported to the morgue for detailed examination. There, the forensic anthropologist sifted through the soil, collected all the bones present and laid them out in their normal anatomical positions."

The role of the anthropologist is to determine the sex, race, height, and approximate age of the deceased, as well as estimate the length of time the individual has been dead.

The forensic odontologist concentrates on examining the teeth, the upper and lower jaws, and the skull. In this case, the deceased's dental records and X-rays taken prior to his death would be compared to X-rays of the excavated skull and jaws.

> "During the search of the accompanying soil, metal detectors located a rusty metallic object. This was tagged and taken to a ballistics expert for analysis. The metal object was identified by the firearms examiner as a .38-caliber bullet."

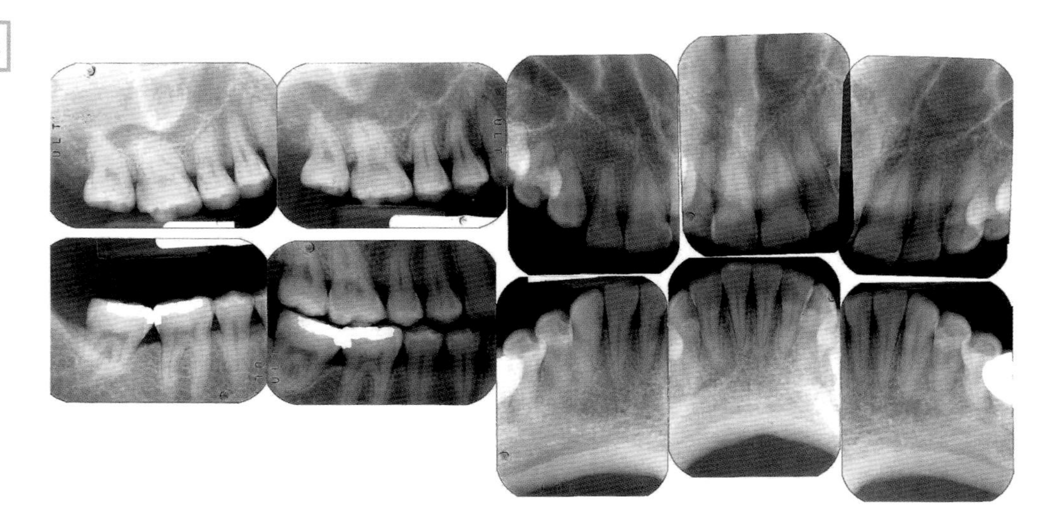

Anti-mortem dental X-rays used by the forensic odontologist to make a positive identification of an unknown victim.

Forensic pathologists also consider nonmedical information when assessing the time of death. This can include whether mail or a newspaper has been picked up, whether lights in a residence are on or off, and whether there is food on the table. Telephone logs are useful, too. Also, clues can be obtained by examining the stomach contents; generally, the average adult stomach will empty within 2–4 hours of eating a meal, and faster in children.

The Death Certificate

It is the duty of the forensic pathologist to fill in the death certificate. This is done after a thorough study of the death scene investigative reports, scene photographs, police reports, hospital records, results of the postmortem examination, and results of the toxicological analysis. After the forensic pathologist completes the death certificate, it is reviewed by the coroner or medical examiner.

The death certificate is a civil-law document, not a medical or scientific document, and it is a public record intended for use by a variety of agencies. Although it is a legal statement of the cause and manner of death, it is not otherwise legally binding on any other agency or individual.

The death certificate will detail the immediate cause of death and any underlying causes; any other significant conditions contributing to death, but not related to the disease or condition causing it; and the manner of death.

The manner of death can fall under one of six categories: natural, accident, suicide, homicide, pending investigation, and cannot be determined. Any physician can issue a natural death certificate, but all deaths of an unnatural cause come under the jurisdiction of the coroner and must be certified by that official.

Natural deaths are caused by naturally occurring diseases, such as cancer, heart disease, and liver disease. Accidental deaths result from behavior or actions that unintentionally end in death. The most frequently encountered accidental deaths are from drug overdoses, motor vehicle incidents, falls, and fires. The deliberate ending of one's life is suicide, and the most common method varies by sex: males frequently choose guns, while females typically take an overdose of pills. Homicide is defined as taking the life of another. Based on this definition, all deaths caused by civilians shooting intruders late at night, all police shootings, and all executions are homicides. The determination of first-degree, second-degree, third-degree, and other types of homicide is made by the court.

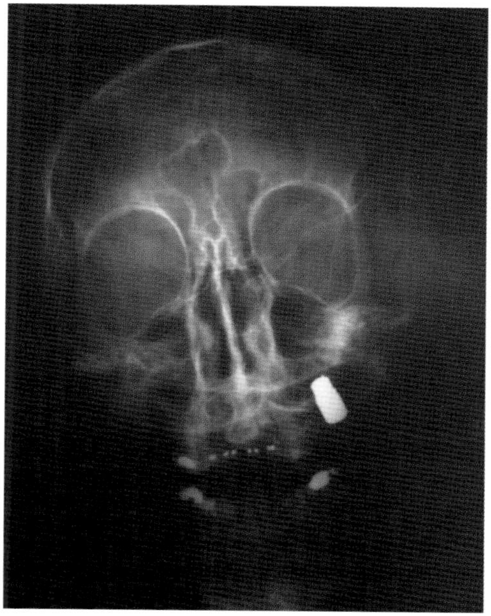

An X-ray of a skull showing a bullet lodged in the left side of the victim.

The cause of death given on the certificate is only an opinion, and represents the best effort of the coroner to describe in a few words the cause based on all available information. In some cases, the most logical determination of the manner of death, and the best estimate of the time and date of injury will be stated when neither investigation nor examination of the deceased provides definitive information. The coroner will use reasonable medical probability when formulating opinions in the same way that clinicians make diagnoses and plans for treatment.

In the United States, the death certificate must be issued within 72 hours, even if the cause of death is unknown; in Canada, the time limit is 48 hours. If the cause is not established with reasonable certainty within that time, the coroner will file a certificate with the cause of death designated as deferred or pending further action. As soon as the cause of death is determined, a supplemental or replacement death certificate will be filed with the appropriate entry.

05 Weapons of the Killer

Murderers use a wide variety of weapons to carry out their crimes—some obvious, like guns and knives; others more subtle, like poisons. Identifying and tracking down those weapons can be challenging for homicide investigators.

The majority of violent deaths involve injuries inflicted by guns and knives. However, a wide variety of items can be used as weapons, including baseball bats, screwdrivers, and rocks. The main goal of a forensic investigation is to determine the cause and manner of death. In addition, the mechanism of death is of interest and relevance. To illustrate different mechanisms of death and their features, four hypothetical crime scenarios are described in this chapter.

Death by Shooting

The basic workings of a gun follow a set sequence. When you pull the trigger, the hammer is released. It strikes the primer in the cartridge and ignites the powder, which expands rapidly and projects the bullet along the barrel. The bullet leaves the end of the barrel (muzzle) at a speed between 810 ft. (247 m) and 4,110 ft. (1,253 m) per second, depending on the type of weapon. The bullet perforates the skin in 50 nanoseconds. At first, the skin stretches as the

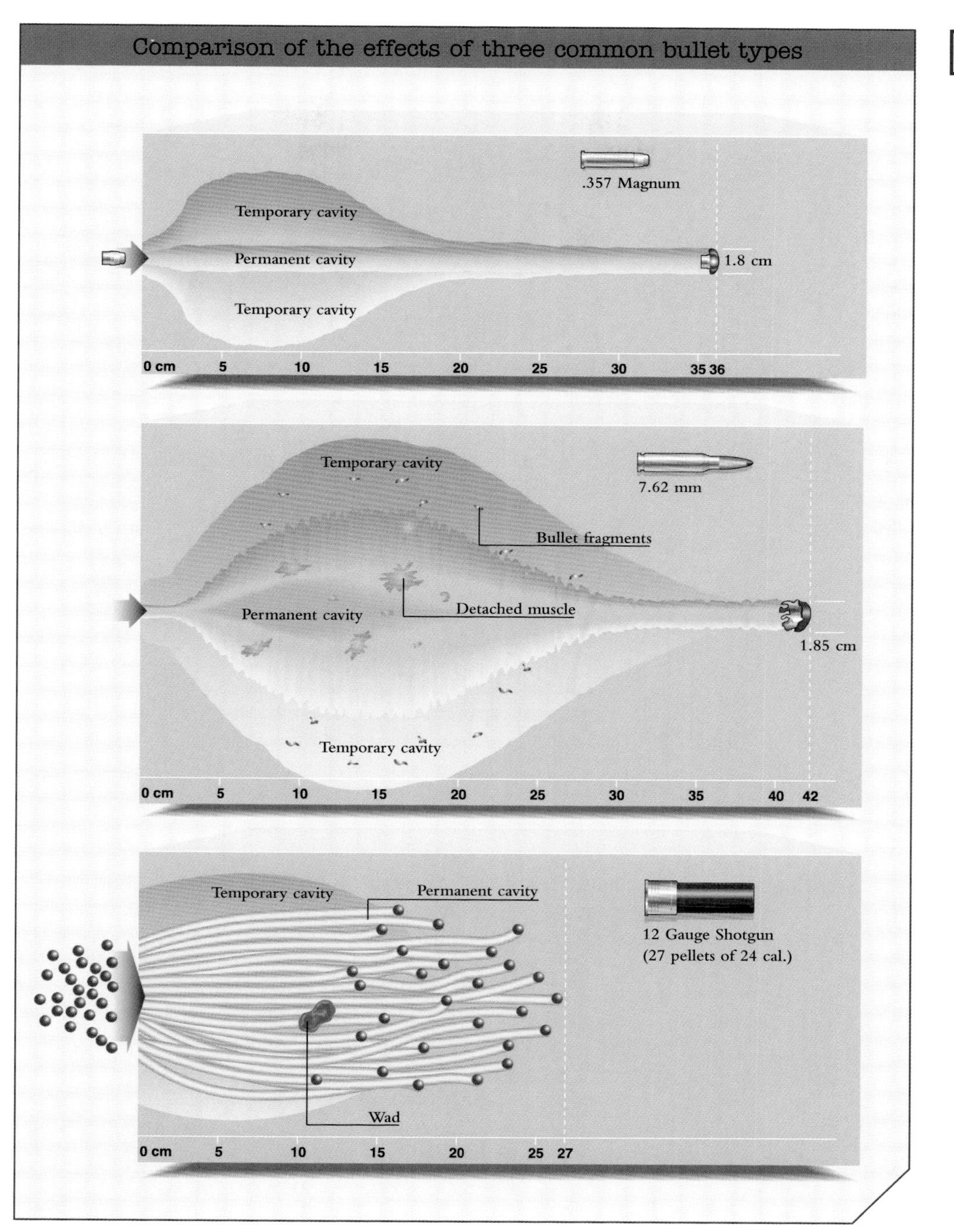
Comparison of the effects of three common bullet types
.357 Magnum
Temporary cavity
Permanent cavity
Temporary cavity
1.8 cm
0 cm 5 10 15 20 25 30 35 36
Temporary cavity
7.62 mm
Bullet fragments
Detached muscle
Permanent cavity
1.85 cm
Temporary cavity
0 cm 5 10 15 20 25 30 35 40 42
Temporary cavity
Permanent cavity
12 Gauge Shotgun
(27 pellets of 24 cal.)
Wad
0 cm 5 10 15 20 25 27

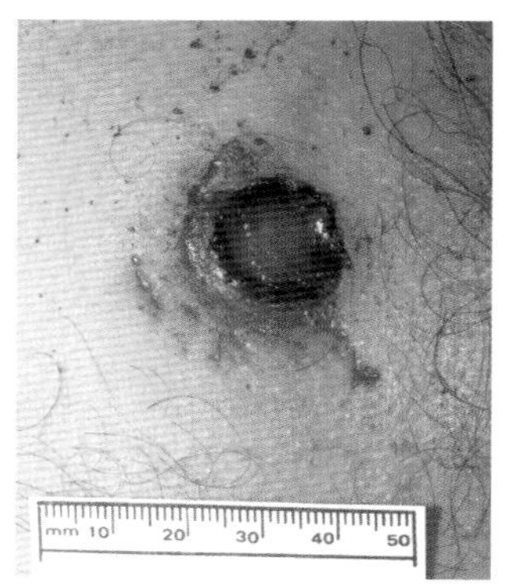

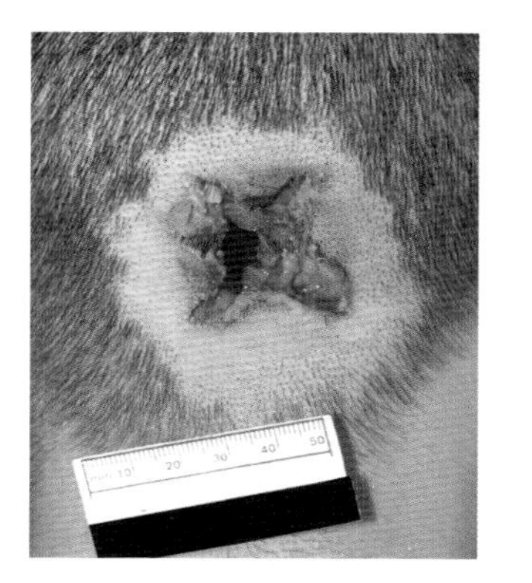

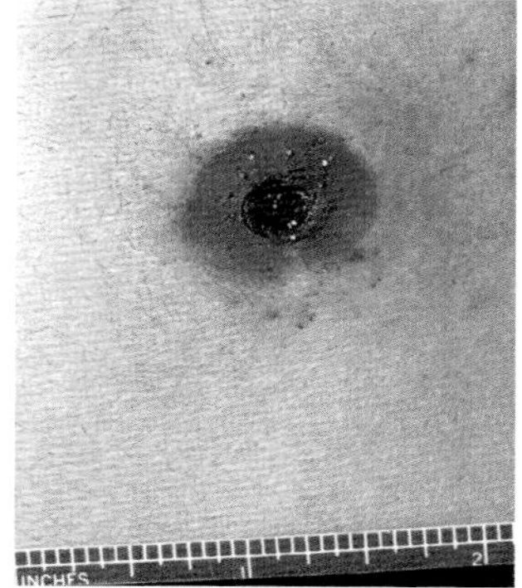

Three gunshot wounds:
The first, from left to right, is a contact wound, the second depicts a stellate, star-shaped, pattern created when a gun is in contact with the skin when fired, and the third shows soot around the wound.

bullet disrupts the epidermal layer, and the energy wave results in lacerations of capillaries and small vessels within the subepidermal layer. The cavity in the tissue is caused as the bullet and gases compressing the neighboring tissue pass through it. This cavity lasts for about three microseconds. After the cavity contracts, it is rimmed with hemorrhaging tissue. This basic sequence can be affected by the type of weapon, the type of ammunition, and the distance from the end of the muzzle to the skin.

Examining gunshot wounds

The examination of gunshot injuries entails identifying the entrance and exit wounds and, for the former, estimating the distance (range) between the victim and the gun when it was discharged. Inspection of the entrance wound may help to determine of one of the four basic ranges: contact, near contact, intermediate, and distant. Clues that help in the assessment of this distance include lacerations, muzzle impacts, soot deposits, and powder stippling around the entrance wound.

If a weapon is fired while in contact with the skin, the resulting injury is called a contact wound. The contact may be tight, loose, at an angle, or incomplete (due to the curvature of the body surface). In tight contact, there may be a patterned abrasion around the entrance wound consistent with the muzzle of the weapon. If the bullet enters over a bony surface, multiple lacerations are usually found around the wound. In such cases, the gases that propel the bullet along the barrel follow it out of the gun, causing the skin to tent up as they encounter the underlying bone and displace the soft tissues around the bullet's path. The skin is pushed upward violently, causing small tears that radiate from the entrance wound. This gives the entrance a star-shaped or stellate appearance. The edges of these lacerations can be pushed together to demonstrate a central defect where the bullet actually entered the skin and tore a piece off. In a contact wound, none of the gas (of which soot is a part) will be deposited on the skin surface. However, soot may be seen on the external surface of the underlying bone.

A near contact wound occurs when the end of the barrel is no more than 0.4 in. (1 cm) away from the skin. In this type of wound, the findings will be similar to those of a loose contact wound, although soot may cover a greater area of skin around the wound. Burned and unburned gunpowder will produce a pattern or tattooing on the skin or clothing.

Intermediate wounds occur where the distance between the end of the barrel and the skin ranges from 0.4–4 in. (1 to 105 cm), depending on the caliber of handgun used, and the type and amount of powder particles in the load (ammunition). Soot deposition can be seen in these types of wounds, but only out to a distance of 8–12 in. (20–30 cm). Powder stippling will be visible around the

A forensic scientist firing a gun in a rifle range during ballistics research. This test is done in order to determine whether the gun was used in a crime. A gun's hammer leaves a distinctive mark on a bullet's cartridge case when it strikes it during firing.

A Day in the Life: DNA Criminalist

Thomas Meyers, Criminalist, Pennsylvania

It is 11:30 P.M. on a cold night in January and the county radio dispatcher calls to notify me of a homicide where the detectives need specialized assistance in processing the crime scene. As a crime-scene team leader, I assess the dispatcher's outline of the incident and make the necessary calls to assemble another criminalist, a latent print examiner, and a forensic photographer.

At the scene, we are briefed by the homicide detectives, then I am escorted through the scene to obtain a firsthand impression.

Shoe prints in the snow approach a broken window, the point of entry into the victim's residence. Blood smears can be seen on the inner windowsill, while large blood droplets trail from the window to a study at the end of a hallway. There are blood smears on a safe in this room; the safe is open, its contents scattered.

The elderly male victim is dressed in pajamas and lying on the floor of the study. There are blood smears on his neck, although no cuts or contusions can be seen.

The scene is documented first through photographs and written notes. A colored wax is sprayed into the shoe prints to increase the contrast for photography and insulate them from the warming effects of the substance used to cast them. Measurements are taken so that scale drawings of the scene can be made later. Glass fragments are removed from the window and placed in a sharps container for comparison later with any found on the suspect or his or her clothing. Blood samples are collected and packaged to prevent contamination.

The area around the victim, his exposed skin, and his clothing are examined for fibers or hairs left by the killer. The smears on his neck are swabbed with a dampened cotton cloth to collect the blood. Meanwhile, the print examiner is busy searching for fingerprints.

When we finally head toward our van, the sun is rising. The second phase of our work is about to begin: logging in the evidence, and slowly and meticulously examining each piece.

As the DNA supervisor, I perform the DNA analysis of any trace evidence from the crime scenes I process. I am also responsible for serological evaluations, such as sperm searching, blood screening prior to DNA, hair examination, and bloodstain pattern determination.

Our laboratory maintains a hard copy of all DNA profiles developed, any profile from a crime scene being reported to the police investigators and the district attorney. When a suspect is apprehended, I compare his or her DNA profile to that recovered from the scene. If a match is found, I send a report to the police investigators, the coroner's office (in death investigations), and the district attorney's office (in death and rape cases).

Sometimes I am called to court to testify—although in 75 percent of cases we are not called to court, such is the overwhelming effectiveness of DNA evidence.

entrance wound. Burned and unburned powder particles will follow the bullet out of the barrel of the gun and begin to spread out. By 0.4 in. (1 cm) there typically will be a spread that is wide enough to allow some particles to impact the skin surface. This impact produces powder stippling, which leaves small reddish brown to yellowish red marks in the skin. Examination of the marks with a dissecting microscope may reveal green or gray-green powder particles in them.

Any injury that is beyond the range of powder stippling is described as a distant wound. In this instance, regardless of whether the shooter is 10 ft. (3 m), or 1,000 ft. (300 m) away from his target, the entrance wound will appear the same. Therefore, such wounds have no feature that allows an accurate determination of range, except to say that it is beyond that weapon's ability to cause powder stippling.

The behavior of the bullet

After being fired, the bullet can do several things. It may miss the intended target. It may pass completely through the victim, either in a vital region (such as the heart) or a non-vital area (such as a hand). It can hit a hard object before reaching the body and fragment, causing multiple entrance wounds, or it may deform and create an irregular entrance wound. It can hit a hard object in the body (bones), typically causing both the bones and bullet to fragment. Normally, these fragments spread out, damaging more organs and tissues as they travel. They can come to rest in the body or produce multiple exit wounds. Finally, the caliber of the bullet, distance of the shot, or type of bullet (for example, hollow-point) may result in the bullet lodging in the body. The path of the bullet inside the body and the damage caused will be determined by the forensic pathologist during the postmortem examination.

Concurrent with the forensic phase of the investigation, homicide detectives conduct their scene investigation. They carry out a detailed search of the house and surrounding property. The doors and windows are examined for signs of a break-in. They talk to family, friends, and employer, inquiring about job stability, money, relationships, any enemies or recent health problems. If a weapon is found, its serial number is checked to verify ownership.

"During the early hours in a suburban community, the silence was disrupted by the sound of a gunshot. The neighbors looked out their windows to see police and emergency vehicles, followed shortly after by the coroner's van.

After the emergency personnel had determined that the victim was beyond medical treatment, the death scene investigation began. The victim, a middle-aged white male was found in the living room lying on his right side. There was a .45-caliber revolver near the body. There appeared to be an entrance wound in his chest and an exit wound in the lower back region. The house was searched for a suicide note, but none was found. The victim and the surrounding scene were photographed, while criminalists searched for any trace evidence, such as fingerprints, blood spatter/droplets, and hair or fibers. After the death investigators had completed their work, the victim's hands were covered by paper bags, and the corpse was placed into a body bag for transportation to the morgue. The weapon was taken to the ballistics laboratory for examination.

Back at the morgue, the victim of the gunshot wound was removed from the cooler and placed on the examination table. Photographs were taken before the clothing was removed and the body was cleaned. The pathologist noted the general appearance of the victim. He was wearing only a blue shirt with a large amount of black material in the middle of the chest area, just below the nipple line, and white shorts. The shirt and shorts were removed and placed on a table covered by white evidence paper. The clothing was allowed to dry and was then wrapped up, labeled, and transferred to the crime laboratory for analysis."

At this stage, the paper bags are removed from the hands and autopsy technicians test for gunshot residue (GSR). When a gun is discharged, GSR (a combination of lead, antimony, tin, and barium) may be transferred to the hands of the person firing the weapon. After the test for GSR, a latent fingerprint technician takes the victim's prints and determines if he or she has a police record and ascertain if there are any outstanding warrants for arrest.

"The examination of the victim revealed an entrance wound in the chest and an exit wound in the lower back. The former was roughly circular, and the skin around it was stained by black soot. In addition, there were signs of powder stippling. The exit wound was oblong in shape.

The external examination also revealed a small, irregular injury, suggestive of another gunshot wound, in the back of the right lower leg, approximately 6 in. (15 cm) above the heel. An X-ray revealed a metal object roughly level with the wound. The site was opened and a deformed bullet was recovered. It was sent to the ballistics laboratory for examination. This raised the question, "Was the victim shot twice?" If so, why did the physical evidence and the witness statements suggest that the victim had been shot only once?

The pathologist was puzzled. The technician, to make conversation, reminded the pathologist that the coming weekend was Easter. At that point, the pathologist asked the technician to pretend that he was praying. He got down on his knees and the pathologist stood above him, pointing at his chest. Using his finger as a gun, he noted that if the shooter had been standing over the victim, the bullet would have entered the chest, exited the back, and lodged in the back of the right leg.

A review of the homicide detective's report indicated that no bullet holes had been found in the walls, ceiling or floor of the living room. Interviews with neighbors confirmed that only one gunshot had been heard. In addition, there were no signs of forced entry. The pathologist requested that X-rays of the trunk be taken. They showed a very small metal fragment in the chest area.

The body was opened by means of a "Y" incision. The sternum and anterior portion of the ribs were removed and the path of the bullet was examined. It was determined that after smashing through

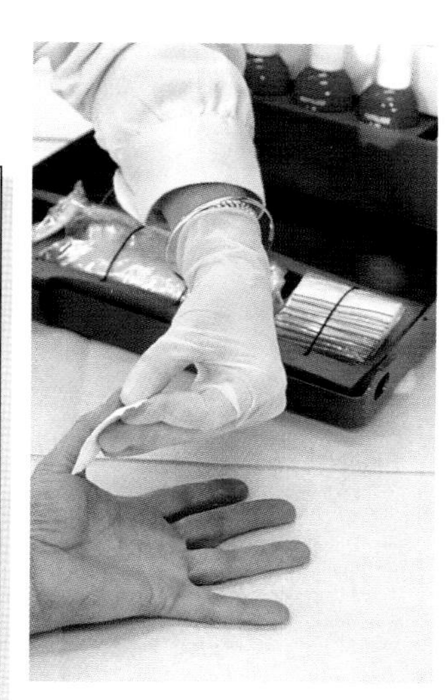

A forensic scientist swabs the hand of a victim to collect any microscopic powder particles that may be present. This test is used to determine whether the deceased had fired a gun and if so, what type of gun.

the sternum, when the small fragment had broken off, the bullet had lacerated the right ventricle of the heart (producing a large amount of blood in the right side of the chest), passed to the right of the spinal column, perforated the soft tissues of the skin of the back, and then exited around the level of the fourth lumbar vertebra. Finally, the bullet had entered the back of the right lower leg, where it became lodged. The path of the bullet was front-to-back, downward and to the right. The remainder of the autopsy was unremarkable.

A call from the homicide detective provided the following information: an investigation by a forensic accountant had revealed that the victim was greatly in debt; the weapon was not registered to the victim; the ballistics laboratory report stated that the bullet recovered from the victim came from the weapon discovered at the scene; no fingerprints were found on the weapon; the test for GSR was negative; and the test firing of the revolver placed it 5 in. (13 cm) away from the victim when it was fired.

Initially, the pathologist had determined that the cause of death was a gunshot wound to the chest, and the manner was pending investigation. Once all the additional information was available, he decided that the victim could not have shot himself and ruled the death to be a homicide. ”

Stab Wounds

Sharp objects cause two main types of injury—incised and stab wounds. An incised wound is longer than it is deep, typically with sharp edges, and is made with a cutting or slashing action. There will be no soft tissue bridges. A stab wound is deeper than its surface length. The edges are usually sharp and straight if a blade is used, but may be patterned if some other instrument is responsible (for example, scissors, an ice pick, or a screwdriver). Abrasions of the edges of the wound are often caused by the handle or hilt of the weapon hitting the skin; they are an indication that the complete length of the blade was inserted into the body. Upon initial external examination, a stab wound may appear trivial.

A knife found at a crime scene and marked for examination by forensics experts.

The number and locations of stab and incised wounds offer some clues to the manner of death. A single wound usually indicates a homicide, especially if it is in the back. While multiple wounds to the arms may suggest a suicide, multiple stab wounds to the chest and head region may point to a crime of passion. A typical suicide involving the neck region is characterized by several superficial cuts, called hesitation marks, and one deeper fatal cut, typically described as a slash, which would display characteristics of both an incised wound and a stab wound. In contrast, a classic homicide wound to the neck consists of one clean, deep cut with no hesitation marks.

Before the autopsy begins, homicide investigators will discuss the case with the pathologist.

“The police were called to the body of a black female, which had been discovered by some boys under cardboard boxes in an alley. The first officer on the scene determined that the victim was beyond medical treatment and sealed off the alleyway with crime scene tape. A short time later, the death investigation team arrived. Photographs of the scene were taken. The boxes that covered the body were removed and placed in large evidence bags. The victim appeared to be in her late 20s and was dressed in revealing clothing. She was face down on her hands and knees. Rolling the body over revealed a moderate amount of dark blood on the ground. There was a stab wound in the left abdomen, and an exhaustive search of the alley

located a bloody knife. The body and knife were transported to the morgue. Meanwhile, the homicide detectives began their investigation by questioning the boys and canvassing the neighborhood.

The detectives told the pathologist that the victim was a known prostitute, and that a witness had seen her talking to a man half an hour before the body was found. Another witness in an apartment above the alley had heard a verbal exchange between a man and a woman. At this stage, there were no suspects.

The victim was placed on the examination table, and a single stab wound was noted on the left abdomen at the level of the navel. It was approximately 2 in. (5 cm) long, running at a 45-degree angle in an upward direction. The edges of the wound were smooth. Internal examination of the organs revealed that the victim's liver had been cut, causing a large amount of hemorrhage. The cause of death was loss of blood, and the manner was homicide.

The right hand showed several incised wounds to the palm and fingers, some superficial and some deep, consistent with defense wounds. ”

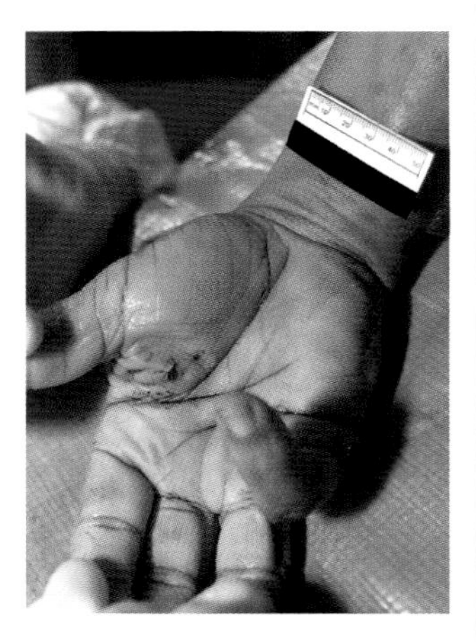

The hand of a victim showing defense wounds.

Defense wounds are caused when the victim attempts to protect the face and chest from attack or attempts to grab the weapon.

“ The pathologist examined the knife used in the murder. It was a 12-in. (30-cm), single-blade kitchen knife and it was covered with blood. The blade appeared intact. The knife was sent to the serology division of the crime lab for the blood to be analyzed. ”

Sometimes, the victim of a knife attack is able to inflict injuries on the attacker by scratching the face or ripping hair from the head. During the autopsy, the hands and fingernails undergo a detailed examination for such evidence. The fingernails are cut and sent to the crime lab, where any materials found under them are collected and submitted for DNA testing.

> “A small amount of skin was discovered under one of the fingernails of the victim, and the genetic profile of the DNA extracted from it was entered into the FBI's Combined DNA Index System (CODIS). A few days after the postmortem examination, the forensic pathologist met with the lead homicide investigator. He told the pathologist that the CODIS system had matched the DNA with a recently released convict, whose mother lived in the area where the girl was killed. Shortly after, the suspected murderer was arrested, and the pathologist made preparations to appear in court to testify to the cause and manner of death.”

Penetrating injuries caused by knives, forks, ice picks, screwdrivers, and broken glass bottles all cause unique and identifiable trauma. While a particular wound cannot be matched to a knife in the way a bullet can be identified as coming from a specific gun, characteristics of the knife can be determined by an examination of the injury.

A knife wound can be described by the following characteristics:

1. Location (in relation to a fixed anatomical mark).
2. Size. The wound is closed with tape and measured while closed. This gives a rough estimate of the blade's size and shape.
3. Orientation. A knife wound can be horizontal, vertical, or oblique. This applies to its orientation on the body surface, not the direction of thrust. For example, an experienced knife fighter will hold the blade with the sharp part facing upward and thrust underhanded; an inexperienced person would hold and thrust the knife in the opposite direction.
4. Borders. A sharp knife will produce a clear separation of tissue. The borders will be regular if the victim does not move, the knife is not twisted, and the tissue is firm. A knife wound with irregular borders, indicates a dull weapon.
5. Edges (the two extremities of the wound that are not borders). These provide clues to the blade's shape and size. A knife with one sharp edge and one squared edge will leave a characteristic wound shape.

A victim's multiple stab wounds are examined by a pathologist to determine the type of weapon used.

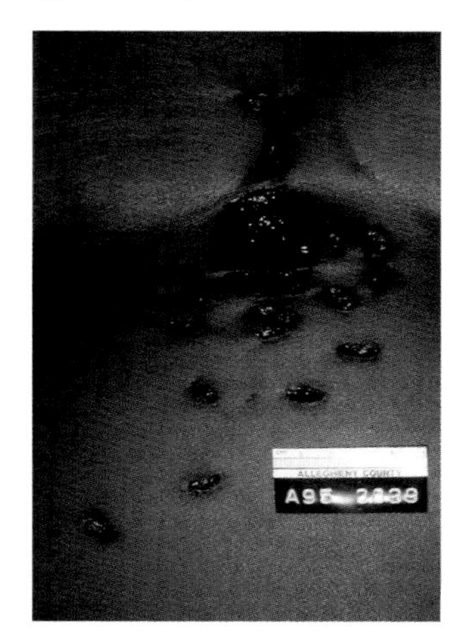

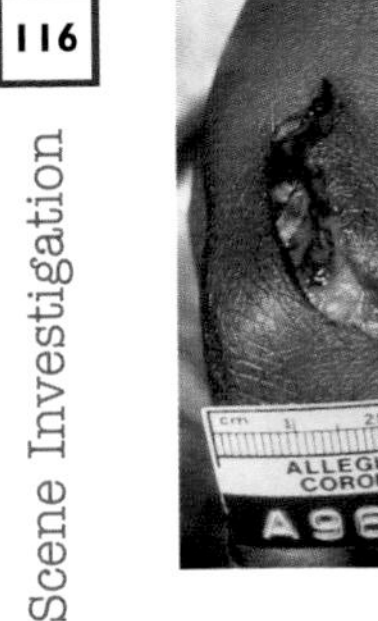

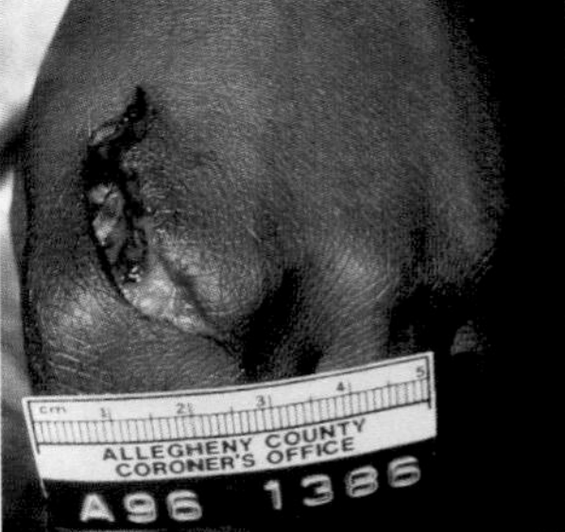

Defense wounds on the top of a victim's hand.

6. Thrust. Once the blade has perforated the skin, it is directed by the thrust, unless twisted by the attacker. A simple thrust allows the direction of the movement to be determined, which can be compared to the attacker's statement.
7. Pattern. The number and location of wounds are significant. A frenzy attack is characterized by multiple stab wounds in a small area with chaotic orientation. Many wounds of this type are non-penetrating. Defense wounds are commonly seen when the victim attempts to grab the knife from the attacker. Deep wounds that contact bone can also leave telltale marks on the bones, which may be matched to the weapon.

The victim's clothing can offer additional clues to the type of knife used. It can provide an outline of the blade due to its increased rigidity compared to skin. It can indicate the position of the victim when stabbed if the bleeding leaves a track in one direction. Fibers from the clothing can adhere to the weapon and be physically matched to provide strong evidence that a particular weapon was involved in the assault.

The pathologist should examine any potential weapon recovered by the investigators, especially if it would produce an unusual pattern, such as scissors or a screwdriver. It is also important to determine if the blade is complete. Sometimes, a part of the knife, typically the tip, will break off inside the body. This fragment can be physically matched to the suspect's knife.

Blunt Force Trauma and Asphyxiation

Blunt force trauma (BFT) occurs when a non-cutting, non-sharp force is applied to the skin and transmitted to the internal organs, exceeding the capacity of the tissues to absorb the energy. Injuries found in BFT cases include contusions, abrasions, lacerations, and fractures. A contusion is caused by hemorrhage into soft tissues and is commonly known as a bruise. If the hemorrhage accumulates and expands, forming a pocket of blood, it is called a hematoma. An abrasion affects only the outer layer of the skin (epidermis), which is shaved off by tangential frictional forces. If a large area is involved, an abrasion may take the form of a brush burn. Laceration refers to the tearing of the skin, usually over a bony prominence, or of the internal organs, which is produced by compression following an impact. A laceration typically has irregular, contused, and undermined borders, with fibers of torn tissue (bridges) clinging between the edges. A fracture is a traumatic injury to bone, in other words, a break or splintering.

"The school principal became concerned because one of his students, a 10-year-old girl, had not shown up for a planned field trip. He was unable to contact her family. He notified the police, who sent a car to her home. As the police officers approached the house, everything appeared normal. They walked to the back of the building and saw a set of footprints in the snow, with drops of what appeared to be blood, leading to a small shed. The officers opened the shed door and saw a middle-aged, white male hanging by the neck from the rafters. He was dead. The officers followed the drops of blood back to the house, where the trail led to the girl's bedroom. She was found face down in her bed, with dried blood on her cheeks. Homicide detectives and two forensic investigation teams were called to the scene.

One investigation team focused on the homicide of the girl, while the other dealt with the suspected suicide of the man. After scene photographs had been taken, the bodies were examined. It appeared that the child had been struck on the head by a blunt object while asleep. A detailed search of her room failed to locate anything that could have caused the injury. Both child and bedding were transported to the morgue.

Prior to the postmortem examination, numerous photographs were taken of the girl's head. Her hair was matted, and dried blood was fixed to her cheeks. The hair was shaved around the impact area to determine the type of object that had inflicted the trauma. When the skin was replaced in its normal position, a crescent-shaped laceration was visible in the left occipital region (back of the head). An incision was made and the skin was peeled forward to expose the underlying skull. The bone was found to be shattered. There was a 1.5-in. (3.8-cm) diameter depressed fracture with several linear fractures radiating from it.

The investigation team dealing with the body of the man found one end of an orange-colored extension cord tied around a rafter and the other around his neck. A box was nearby; it appeared that he had stepped onto the box and then kicked it away. A search of the shed unearthed a bloody hammer wrapped in an old rag. The shape of its head was consistent with the injuries noted in the girl's skin and skull.

One death scene investigator cut the cord from the beam, while the other supported the victim and lowered him into a body bag. They were careful not to cut the knot or remove the cord from the neck. ”

The majority of individuals who hang themselves actually die from asphyxia rather than a severed vertebral column and spinal cord. Either the knot is misplaced or the drop is too short.

There are two basic forms of asphyxia: intrinsic and extrinsic. The former occurs when some type of physical barrier prevents the intake of oxygen, such as aspirated food, vomit, or dentures. Within this category is functional asphyxiation, which is caused by the aspiration of a gas such as carbon monoxide, carbon dioxide, or cyanide. In various ways, these stop the hemoglobin molecules in the blood from picking up and transporting oxygen. Extrinsic asphyxiation occurs when a force is applied to the chest, preventing it from expanding. Hanging is included in this category.

The purest definition of hanging is when the body's weight, acting on a constricting ligature, causes compression of the airway, blood vessels or both. In a complete hanging, the body is suspended off the ground. In contrast, the body in an incomplete hanging is in contact with the ground. While the majority of hangings are complete, it is a myth that the body has to be suspended to die from asphyxiation. This can occur even if the victim is on his knees. Regardless of the mechanism, the process of asphyxia generally follows three phases. In phase one, hypoxemia (lack of oxygen in the blood), hypercarbia (increase of carbon dioxide in the blood), and acidosis (increase in blood acidity) develop rapidly, with increased heart and respiratory rates and struggling. The more the victim struggles, the quicker he moves into phase two. This is marked by congestion of the face and neck above the level of compression, leading to hemorrhages in the membranes of the eyes and of the skin among other tissues.

A Day in the Life:
Forensic Knot Analyst

John J. Van Tassel, Royal Canadian Mounted Police, Vancouver, Canada

At most crime scenes, the offender will leave behind some type of evidence. This may not always be visible, such as latent fingerprints, or it may be very obvious, such as a discarded weapon. One piece of evidence often found, either on the victim or nearby, is a ligature. Ligatures are used to bind, gag, garrote, hobble tie, and hang victims. My job is to extract as much information as I can from any ligature recovered.

My typical day would begin by attending the autopsy of the victim. Whenever possible, ligatures are removed from a victim in the presence of, or under the direction of the pathologist. Before cutting a ligature, the site of the cut is wrapped with surgical tape printed with matching pairs of numbers or letters, then the cut is made between these pairs of figures. The tape prevents the cut ends from fraying, while the numbers or letters allow their reassembly during analysis, which is done in the forensic laboratory.

In the lab, the first thing I record is the ligature material. It might be an item of the criminal's clothing, such as a belt, or a piece of rope carried by him or her to the scene, or something found at the scene itself. Materials I have come across include: rope and cordage; belts, neckties, brassieres, pants and jackets; and metal cable, chains, and electrical cords. Sometimes the material gives a clue to the criminal's occupation or hobby. A heavily waxed rope lariat, for example, suggests someone in ranching or farming, while monofilament line points to a sport or commercial fisherman. Since the material was handled by the criminal, it may contain DNA or other trace evidence, such as hairs and fibers. For this reason, I cooperate closely with other forensic specialists.

Knotted ligatures are normally used to bind the hands and feet, as well as to gag or garrote a victim of a homicide. There are other instances, however, when bound hands and feet are found: in some suicides, for example, and cases where unconventional sexual activity has caused the death of the victim (autoerotic) or one of the participants (sadomasochism). By studying and identifying the knots used, I can determine the skill level required to tie them, their sequencing, whether the victim could have tied them himself or herself, whether they were used or tied incorrectly, whether there may have been more than one offender, and possibly the occupation or hobby of the offender, since many make use of specific knots. As I work, I photograph my actions and any changes I might make to the evidence (by untying a knot, for example). All this information can help point to a potential suspect or suspects.

Sometimes I am called to the crime scene itself, usually when a victim has been tied in a particularly complex way or when removing a victim may compromise the knotted evidence. At other times, I may simply be sent a package containing the ligature, together with an outline of the crime and any questions the investigative team might want answered.

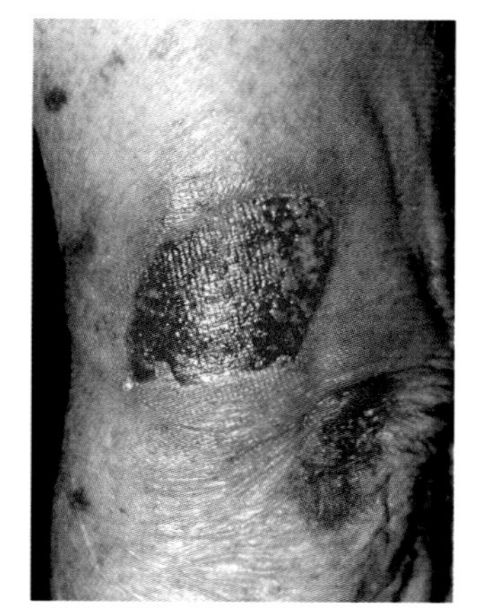

Skin loss on an elderly woman's arm after a trauma. Even minor traumas cause large amounts of damage on frail victims.

The third phase is highlighted by seizures, sweating, loss of consciousness, regurgitation of the stomach contents into the esophagus, loss of sphincter control, slowed heart rate, and finally, death. It should be noted that death could occur during any of the three phases.

> "At the morgue, the male victim was placed on the autopsy table, maintaining the position of the cord. After noting the latter, the pathologist removed the cord to reveal an underlying indented ligature mark around the neck consistent with hanging."

In cases of hanging, the most common knot encountered is the simple lasso or running noose. This is followed by the double overhand or granny knot. Rarely does anyone use the true hangman's knot, consisting of seven or eight coils.

> "After completing both autopsies, the forensic pathologist met with the homicide detectives. Their investigation had discovered that the parents of the girl had been in a violent divorce proceeding, and the father had just learned that the judge had granted complete custody of the child to her mother. Apparently, he had gone to the child's bedroom and delivered a single blow to the back of her head with the hammer, then carried the weapon to the shed. There, he had hanged himself."

Poisoning

Poisoning, the introduction into the body of an amount of a substance incompatible with life, can take many forms. Death by poisoning can be an accident, a suicide, or a homicide. For example, the death of an elderly woman with dementia, who takes three times the recommended dosage of her hypertensive medication, is an unfortunate accident. A middle-aged man found dead with a needle in his forearm also is likely to be a victim of an accidental drug overdose. The young man with financial problems found in his car in a closed garage, with a hose leading from the exhaust pipe into the car, could be a suicide by carbon monoxide poisoning. The deliberate administration of a chemical substance in lethal levels to another individual is a homicide.

In most cases of death by poisoning, the circumstances surrounding the death and the past medical history allow a fairly straightforward conclusion to be drawn as to the cause and manner of death. However, not all cases may be as straightforward as they initially appear. Consider the following case:

> "A 10-year-old girl went to visit her grandmother, who lived alone after the death of her husband two years before. During her visit, the young girl said that she was thirsty, and the grandmother told her to have a glass of the "Half & Half" grapefruit drink in the refrigerator. The girl poured herself a glass and put the remaining soda back into the refrigerator.
>
> Shortly after she had finished the drink, the girl's parents arrived to take her home. During the trip, the girl began to complain of stomach cramps. When the family arrived at their home, she was quite ill and went to bed. Within the hour, she was vomiting constantly and in extreme pain, so her parents took her to the local hospital emergency room. At first, the medical staff suspected a ruptured appendix and drew blood for testing. However, the results were negative for infection. Then the girl's condition worsened, and she became unconscious and died.
>
> The body was taken to the coroner's office for an autopsy because the death was unexplained and sudden. The postmortem examination failed to show any anatomical or pathological evidence to explain the death. Blood, bile, urine, and eye fluid were drawn for toxicological testing, which indicated the presence of a lethal level of strychnine."

At this point in the investigation, the cause of death was known, but determination of the manner of death was still pending. Could this be a homicide? Was the sweet old lady capable of murder? Detectives began to question the

Margie Velma Barfield, from North Carolina, confessed to the poisoning murder of her mother and three others in 1984. Hers was the first execution of a woman in 22 years.

grandmother, who was now the main suspect in a possible homicide. She told the police that the only thing the girl had consumed was the "Half & Half" soda, the balance of which was still in the refrigerator. She also informed them that it was a leftover from when her husband had died. He had suffered from stomach cancer, and toward the end, "Half & Half" was about the only thing he could swallow.

The investigators took the bottle of soft drink to the forensic lab, where an examination revealed an accumulation of a dried substance around its mouth and neck. This residue and the remaining liquid in the bottle were analyzed. Both contained high levels of strychnine—lethal levels if consumed. The detectives contacted the manufacturer and, through the numbers embossed in the glass, were able to determine that the soda had been made and bottled two years before, and that it had been sold to a distributor in the area where the grandfather had lived. The manufacturer also reported that there had been

Case file: Joann Curley

In August 1991, Pennsylvania electrician Robert Curley was admitted to the hospital with a mysterious illness. He had leg pains, numbness, nausea, and burning in his hands and feet. His condition improved, however, and he was discharged on August 29. Nine days later, he was back and getting worse. On September 22, he recovered sufficiently to see his wife, brother, and sister, but that night deteriorated and continued to do so until his wife agreed to the withdrawal of life support on September 27. By then, doctors had discovered high levels of thallium in his system.

Curley had been working in a building that contained jars of thallium salts and, at first, it was thought that he had been accidentally exposed to the poison there. Tests showed that this was not possible, however. Moreover, the levels of thallium in his body could only have come from ingesting it through eating or drinking.

One theory was that a fellow worker may have played a misguided prank on Curley by adding the thallium to his iced tea, thinking it was the similar-sounding Valium tranquilizer. Investigators interviewed his coworkers, but with no positive result. Since Curley had sought treatment, the case was ruled a homicide.

Years passed without any real progress. In 1994, however, a forensic toxicologist offered

no similar incidents, so it was unlikely that this was a case of large-scale product tampering.

Could it be the grandmother?

Again, the investigators questioned the grandmother, who told them that her husband had often purchased the soft drink at a small general-purpose store in a nearby town, but that she knew nothing about the strychnine. The detectives questioned the store's owner, who said that he remembered the deceased man, who had often purchased one or two cases of "Half & Half" at a time. He also recalled that just before the old man died, he had bought two packages of strychnine-based rodent poison, saying that he had a serious problem with rats in his barn.

The doctor who pronounced the grandfather dead was consulted. He verified that the man had had stomach cancer, which caused his death. The only thing out of the ordinary, as far as the doctor was concerned, was that he had not expected the death so soon. He had thought the man would live for another year, perhaps more. The body did not

to produce a timeline indicating the peaks of thallium ingestion, which would give a clue as to how Curley had been exposed to the poison. The body was exhumed and samples taken of hair, nails, skin, and other tissues where thallium would be deposited. The analysis showed that the thallium levels in Curley's body began to rise in late 1990, then dropped, then rose, and continued seesawing throughout 1991 until a final massive increase caused his death. It was puzzling that this had occurred even while he was in the hospital.

It became clear that Curley must have ingested the final dose on September 22, when he received visits from his family, and the finger of suspicion pointed at his wife, Joann. On his death, she had collected over $296,000 from an insurance policy. When investigators studied the peaks of thallium ingestion, they discovered that only Joann Curley had access to her husband on every occasion.

On July 17, 1997, Joann Curley admitted to murdering her husband with rat poison. The Curleys' marriage had been only a few months old when Joann began poisoning Robert because it had not turned out the way she thought it would. When asked why she didn't simply divorce him, she replied that she wanted the insurance money. She received a sentence of 10–20 years.

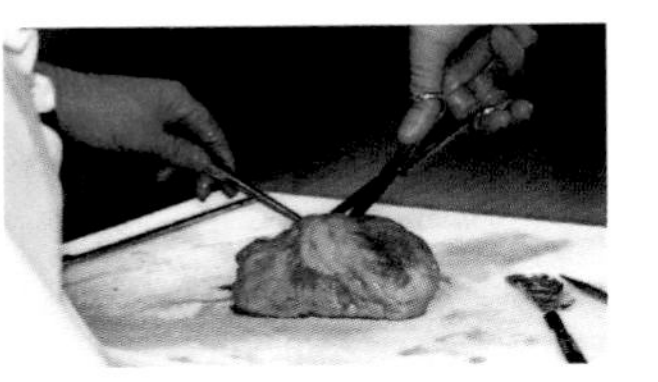

ABOVE A forensic pathologist examines a heart during an autopsy to ascertain whether any toxins were involved in the victim's death.

undergo an autopsy or toxicology analysis. When questioned further, the doctor said that he had been treating the man's pain, which had been growing worse, and that the man had started to have violent fits of chronic vomiting. These were symptoms of his type of cancer. The only thing the man could keep down were fluids such as soft drinks.

When all the bits of information from the varying sources were put together, the chain of events leading to the girl's death emerged. It was determined that the grandfather had purchased the strychnine, and when he could no longer stand the pain, he had mixed it with the soft drink and committed suicide. Because the symptoms of the poisoning mimicked those of his cancer, no one questioned that his death was anything else but natural. The unsuspecting grandmother had placed the remaining soft drink in the refrigerator, and when her granddaughter consumed it, she died.

BELOW This MRI scan of a man's brain shows great damage caused by long-term poisoning.

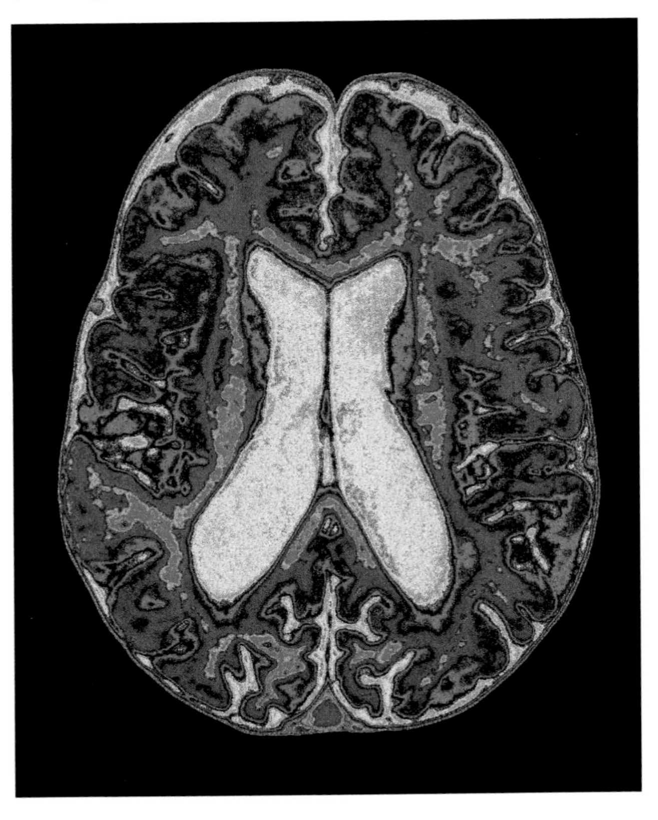

When the answer is not always obvious

The previous stories are just a few examples of methods that individuals typically use to kill others or themselves. However, anything from a roll of duct tape to a 1,200-lb. (544-kg) vehicle can be used as a weapon. In addition, not all weapons leave trace evidence that can be seen. Consider the case of an elderly man who opens his door to see someone pointing a shotgun at his head. Terrified, he suddenly dies from a heart attack. This would be a homicide, even though the victim has no visible trauma to indicate a wrongful death. Thus, forensic scientists and homicide detectives investigating a suspicious death must always consider less obvious means and scenarios.

A Day in the Life:
Forensic Barefoot Morphologist

Robert B. Kennedy, Forensic Identification Research Services, Ottawa, Canada

A typical day in FIRS (Forensic Identification Research Services) encompasses research in barefoot morphology, preparing for and giving lectures and courses relating to physical evidence, and preparing forensic evidence for court presentation.

Forensic barefoot morphology is a method by which the weight-bearing areas of the bottom of a bare foot, excluding ridge detail, are compared to an impression left in some sort of medium (mud, blood, insole of a shoe, for example) found at a scene of a crime, for the purposes of excluding or linking suspects to the crime. Linking a suspect to footwear found at a scene has been done historically in Canadian courts as far back as 1948 and is treated as any other physical match determination, incorporating the similar methodologies and scientific principles as the other physical match fields. It is only recently, in the last 14 years, with the advent of computerized databases and statistical analysis, that the idea of footprints being unique has been seriously researched.

Human beings can take thousands of steps a day, wearing footwear for many of those steps throughout a lifetime. As the foot generates heat and sweat and puts pressure on the inside of the footwear, the footwear responds by stretching and molding to the foot. Areas of the insoles in the footwear become compressed and stained, such as the toe pads, ball area, and heel. These areas, as well as accidental characteristics that are made during wear (such as blister or callous formation) are compared and characterized during this forensic analysis.

When asked to do an examination of a typical homicide case, I first obtain the barefoot impressions from the crime scene, whether they are found in blood or on the insole of a pair of shoes. The inked barefoot impressions from any suspect are obtained along with the shoes he or she may have been wearing at the time of arrest. Proper photographs must be taken of the top, sides, and bottom of the bare foot from the suspect.

When the exhibits are received at this office, they are marked with my initials, dated, and the packages are opened. All exhibits are separated and photographed using a scale so the impression can be photographically enlarged 1:1. The footwear is cut apart and all debris, hair, fibers, and dirt are collected and packaged for possible further examination. The barefoot impressions found inside the footwear are also photographed. The impressions from the crime scene are compared to the shapes, placements, and contours of the barefoot impressions from the suspect, which can either include or exclude that individual as having made the impression.

A report is then drawn up stating my conclusions from the examination, and all the evidence is secured in case I am required to give testimony in a court of law about my findings.

06 Psychology of Crime

What drives someone to commit murder? Or any other crime? The motive for a criminal act can be simple or complex, and may not always be as it first appears. Finding and understanding motive are essential steps in bringing the criminal to court.

Motive is a reflection of intent, criminal or otherwise. Criminal liability requires criminal intent. Therefore, arguing the motives of actions is a prerequisite for presenting criminal cases to judges and juries.

Assessing motive is not the responsibility of forensic professionals. Psychiatrists who evaluate whether someone is criminally insane, for example, do not necessarily have an opinion on why a certain crime happened, only whether the suspect was predisposed to carry it out or appreciated right from wrong.

Invariably, however, juries want to consider why a crime was committed while deliberating on the responsibility for it. Whether or not motive has legal significance, it represents a human aspect of each case. When psychiatrists testify in court, particularly when a crime is unthinkable, jurors may attach more credibility to testimony that educates them about the defendant's motive. Forensic psychiatrists and psychologists are uniquely

John Allen Muhammad on trial in 2003. Muhammad was convicted in the "Washington, D.C. sniper" cases.

FBI's organized/disorganized offender dichotomy

The primary foundation of the FBI's system lies within the organized and disorganized offender dichotomy. The table below illustrates some differences the FBI draws on between these two types of offender.

Organized Offender	Disorganized Offender
Above- to above-average intelligence	Below-average intelligence
Socially competent	Socially inadequate
Skilled work preferred	Unskilled work
Sexually competent	Sexually incompetent
High birth order status	Low birth order status
Father's work stable	Father's work unstable
Inconsistent childhood discipline	Harsh discipline as a child
Controlled mood during crime	Anxious mood during crime
Use of alcohol with crime	Minimal use of alcohol
Precipitating situational stress	Minimal situational stress
Living with a partner	Living alone
Follows crime in news media	Minimal interest in news media

Psychologist Maisha Hamilton Bennett speaks to the Chicago press after consulting on the extraordinary case of two children, ages seven and eight, accused of the murder of Ryan Harris, age eleven.

Case file: George Metesky

Over a 16-year period in the 1940s and 1950s, New York was subjected to a series of bombings that provided few leads for the police to pursue. In some cases, the bomber left threatening notes, while he mailed other letters to the police and the local Consolidated Edison power company.

In desperation, the police turned to a criminal psychiatrist, Dr. James Brussel, in the hope that he could shed some light on the type of person responsible. After reviewing the case, Brussel produced a profile that proved remarkably accurate. Among his deductions were that the bomber was likely to be male,

suited—given the specific qualities of forensic examinations for criminal responsibility or presentencing issues—to probe motive. In doing so, they review the entire case file of police investigation notes, autopsy results, physical evidence reports, past medical and psychological records, school and employment records, and drug and toxicology reports, as well as a defendant's financial transactions, e-mail correspondence, phone calls, and Web activity. Moreover, the psychiatric examiner learns about important aspects and events of the criminal's life by interviewing eyewitnesses, close friends, relatives, enemies, acquaintances, and the actual defendant.

The comprehensive nature of forensic psychiatric examination provides the criminalist with a complete appreciation of the individual—how this person developed up to the time of the crime, and why he or she made certain criminal choices. From this painstaking turning over and fitting together of puzzle pieces—and the ongoing quest for such pieces—comes the beginning of an understanding of motive.

middle-aged, foreign-born, paranoid, and probably living in nearby Connecticut.

The wording of the letters suggested that English was not the bomber's first language. There was a large Slav community in nearby Connecticut, and some of the letters had been mailed from that region. The content of the letters indicated that the writer felt the whole world was against him, and such paranoia is usually strongest around the age of 35; since the bombings had been going on for 16 years, it was likely that he was in his early fifties. Finally, the letters made it clear that the bomber had a particular grudge against Consolidated Edison, so he could be an ex-employee.

Armed with the profile and the records of the power company, the police tracked down George Metesky in Waterbury, Connecticut. He had been injured while working for a subsidiary of the company, but his claim for compensation had been denied. At the time, he had written letters threatening retribution, and the bombing campaign was the result. He was tried and found guilty but also found insane. He was sent to the Matteawan asylum for the criminally insane, where he remained until 1973.

At the asylum he was unresponsive to treatment, believing the psychiatrists were part of the conspiracy against him. He died in 1994, at age 90.

Primary, Secondary, and Tertiary Motive

"Why did the criminal commit the crime?" For the forensic psychiatrist, this obvious question is supplemented by: "Why did the criminal do it at that time, rather than sooner or later?" Motive may appear obvious in some cases, but experience in psychiatric examination reveals that often there is more to a story. If concrete matters such as guilt cannot be presumed, neither can motive.

Motives are often complex or multilayered. A criminal may have a primary motive, secondary motive, and tertiary motive for carrying out a crime.

> "Bashra was an unassuming housewife in Gaza, under the rule of the Palestinian Authority. Her husband was a leading member of the terrorist group Hamas, which is dedicated to killing as many Israeli Jews as possible. Thus, when Bashra went to a checkpoint and blew herself up when Israelis answered her contrived call for medical assistance, her allegiance to her husband appeared obvious. She left behind a videotape, as is the custom for Hamas terrorists, posing in military gear with her child and an automatic weapon."

Was Bashra's motive terrorist mass homicide? Closer study of the case revealed that her family did not participate in demonstrative mourning, as is the custom when Gaza villagers gather to celebrate such deaths. Instead, the family was quiet and unobtrusive, some even absent. Subsequently, it emerged that Bashra's husband had lost his love for her, so she had volunteered for the mission to win him back. Was her motive also romantic? Later, it was learned that Bashra had entered into an affair with a local man but had broken it off. When her family became aware of this, they had shunned her, reportedly leaving her desperately sad and depressed. Was she, then, suicidal in her motivation? Vengeful toward the family that had shunned her? Endeavoring to restore her honor by terrorism, the most decorated achievement within her culture?

Only a thorough investigation, including detailed interviews of witnesses and confidantes, and study of personal writings and other communications made at the time that the criminal decision was made, enables an examiner to

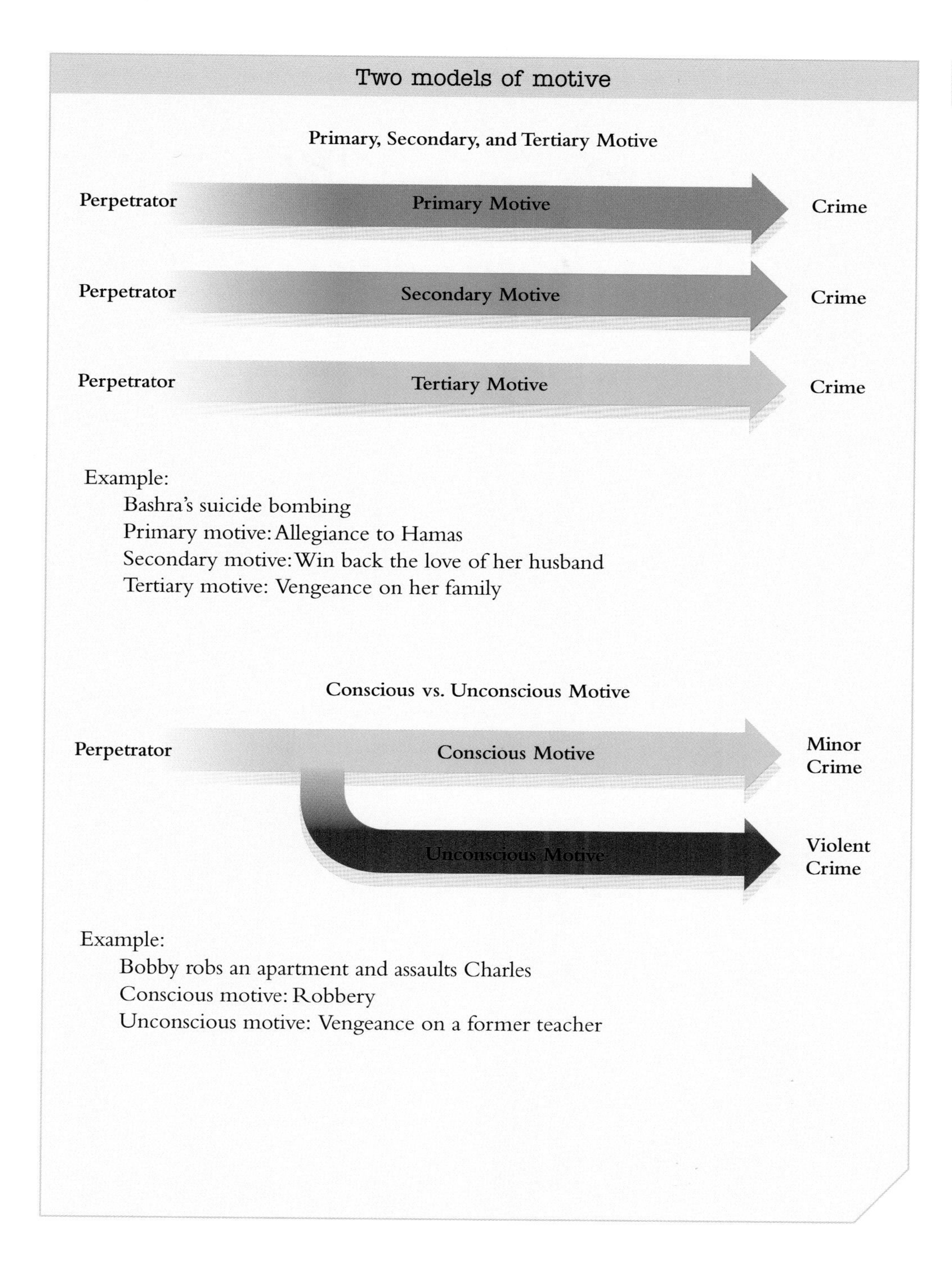
Two models of motive
Primary, Secondary, and Tertiary Motive
Perpetrator
Primary Motive
Crime
Perpetrator
Secondary Motive
Crime
Perpetrator
Tertiary Motive
Crime
Example:
Bashra's suicide bombing
Primary motive: Allegiance to Hamas
Secondary motive: Win back the love of her husband
Tertiary motive: Vengeance on her family
Conscious vs. Unconscious Motive
Perpetrator
Conscious Motive
Minor Crime
Unconscious Motive
Violent Crime
Example:
Bobby robs an apartment and assaults Charles
Conscious motive: Robbery
Unconscious motive: Vengeance on a former teacher

determine which motive or motives may be considered primary, and which motives had lesser influence. Even with a living defendant who can be interviewed, that person's understanding of his or her own motives may change. A person reflecting on criminal behavior while in custody may reach an understanding that bears little resemblance to the motivation at the time of the crime.

> "Jimmy had been working at the Burger King for six months when he decided to rob the place. He studied the shift changes, noted how the manager secured the safe, and learned when he would be able to gain access without getting caught. At the time, he felt that he had done it because it was easy, he resented the management, and he wanted the money.
>
> Jimmy hauled away $6,500. He bought his girlfriend Shaneeka a fur coat before taking her away to the Bahamas for a weekend. They had a blast. But

Case file:

Paul Bernardo and Karla Homolka

To their neighbors in the Canadian town of Port Dalhousie, Paul Bernardo and Karla Homolka were the perfect couple. They had met in late 1987, when he was 23 and she was 17, but by that time, Bernardo had already begun a career as a violent serial rapist in the Scarborough area of Toronto. He continued to rape while he dated Homolka, who gradually fell under his dominant personality.

Desperate to catch the rapist, the police called in profilers. Having studied the available evidence, they estimated that he was in his early 20s and living with his parents.

Shaneeka dumped Jimmy weeks later, taking up with Rashaun, a local drug dealer. Reflecting on a woman lost, Jimmy realized that the reason he had carried out the robbery in the first place was because Shaneeka was a high-maintenance woman whom he knew he would have to ply with expensive gifts to keep. It was all about the woman; not about the money at all. ”

Primary motive reflects the driving force of the criminal's thinking and planning, compelling him or her to disregard the potential for fines, imprisonment, or loss of standing.

Conscious vs. Unconscious Motive

The unconscious has great relevance in criminal motivation. The idea for a criminal action germinates and elaborates into a chosen target, chosen accomplices, or chosen modus operandi, on a chosen date. These features can indicate the influence of conscious and unconscious motivation.

Chillingly, they predicted that he would become progressively more violent and, ultimately, would murder. Although the profile could not identify the rapist, the police did have an artist's impression, and this was recognized by an acquaintance of Bernardo. He was questioned and gave samples for DNA testing, but due to a tragic oversight they were not processed immediately.

Meanwhile, Homolka had fallen so deeply under Bernardo's control that she helped him drug and rape her 15-year-old sister, who died later from accidentally choking on her own vomit. Subsequently, she assisted Bernardo in the abduction, sexual assault, and murder of two more teenage girls.

By early 1993, Homolka had married Bernardo, but had grown sick of his violence and confessed to the police. In need of more evidence to ensure a conviction, they turned once again to profilers, who revealed that sadistic rapists usually kept precise records of their crimes so that they could relive them later, but that these would be well concealed. Homolka confirmed that Bernardo had videotaped their victims, but she did not know where the tapes were kept. A subsequent detailed search of the couple's home unearthed the tapes behind a bathroom light fixture.

Bernardo received a life sentence for the crimes, while Homolka was given 12 years under the terms of a plea bargain.

> “Bobby planned and carried out a robbery, breaking into an apartment that seemed vulnerable. As he was rifling through drawers, he was surprised by Charles, the 53-year-old resident. Charles showed no resistance when Bobby demanded that he go into the bathroom, where he was to remain until Bobby had completed the robbery and left. Twenty-seven-year-old Bobby, 6 ft. 4 in. tall and a fit 220 lb., had nothing to fear from Charles, a 5-ft. 8-in., 150-lb., beer-bellied couch potato. Still, Bobby attacked him and punched him a number of times around the face. Arrested for a violent crime that was totally unnecessary, Bobby conceded that he had not planned to victimize Charles only to rob the home of portable valuables. Ultimately, he reflected that there was something familiar about Charles' voice—it reminded him of a teacher who had suspended him when he was in high school. But he had not recognized this motive until much later.”

The home of Lieutenant Colonel Robert Workman, who was murdered on his doorstep, in Hertfordshire, United Kingdom in 2004. The motive for the murder of this 86-year-old widower continues to baffle investigators.

The complexity of criminal motivation should not be seen as an excuse for bad behavior. However, respect should be given to the individual's particular circumstances. Seeing each defendant as a person enhances justice, whether it is at a trial, sentencing, or a later parole hearing. Understanding primary vs. secondary motives and conscious vs. unconscious motives invariably improves a jury's ability to determine the severity of a criminal act, such as manslaughter vs. murder.

Motive and Gain

Understanding the nature and extent of gain is helpful to making sense of motive. Primary gain is the most obvious benefit of a crime to a criminal. Secondary gain is less obvious, but still important.

> “A robber conceived a bizarre plot to trick a pizza delivery man into wearing a bomb around his neck and entering a bank to demand money. The robbery was successful, but police stopped the delivery man when he attempted to drop the money off as directed by the plotter. Television crews arrived on the scene. In the ensuing standoff, the plotter became aware that the proceedings were being televised and detonated the device, instantly killing his unwilling accomplice.”

Bank robbery is a high-risk, high-yield crime. The criminal wanted a lot of money, but was unwilling to risk capture. Thus, he engaged an accomplice to confront the teller. Unusual to this crime, however, was the use of a bomb and the way in which it was rigged, along with a complex set of instructions for the delivery man to follow. Maybe he had planned to kill him anyway. Perhaps the secondary motive was to orchestrate a crime of spectacular destruction for attention—as bombers are apt to do.

Gain can be both material and psychological, as the previous example illustrates. Clearly, the primary motive was material because the criminal did not choose a less-protected, easier target; his selection of the riskiest target reflects his desire to secure as profitable a haul as possible. Attention seeking was not the primary motive because the bomb went off only after the accomplice had been detained.

A Day in the Life: Criminal Profiler

Dean A. Wideman, Consulting Forensic Scientist and Criminal Profiler, Texas

As a consulting forensic scientist and criminal profiler, most days I find myself studying violent and sexual crimes—as many as a hundred a year, of which 10–15 require offender profiling. I may be contacted by law enforcement agencies, victims' families, victim support groups or legal practitioners about unsolved cases when there are no suspects and no investigative leads; such cases can be months or even years old. I may also be asked to assist in active investigations when serial offenders are sought.

For every case, I examine, analyze, and assess all components of the crime, including the characteristics of the victim, crime scene ,and offender (if known), and the forensic evidence to generate a behavioral profile of the offender; suggest additional forensic laboratory tests and analyses; derive investigative strategies; and provide a geo-forensic pattern analysis in an attempt to isolate an offender's emerging crime pattern, and identify any links between the offender and victim(s). Depending on the amount of available information, this process can take several days, weeks, or months.

Information that I typically receive for examination and review can include forensic photographs, police reports, autopsy reports, victim statements, crime lab reports, family member statements, medical reports, and crime scene notes and sketches. In some cases, I will also reexamine and analyze the physical evidence recovered from the crime scene.

It is very important that I speak to everyone involved in the case, such as the police officers, forensic scientists, and victim's family members. Information obtained from these sources can be crucial to my final analysis and assessment, since the more detailed the information I receive, the more specific I can be in my conclusions, recommendations, and behavioral profile.

Whenever possible, I visit the crime scene. This can be very useful, especially if it has not been disturbed or altered since the incident. Even when a crime is months or even years old and the scene has not been preserved, it is still worthwhile visiting the area to gain an understanding of the local geography and demographics, and the offender's actions.

In creating an offender profile, I rely on my casework experience and hard research. Part of the process involves statistical modeling of data and information gathered on known offenders, but that is only part of the profiling process. A complete offender profile is based on a culmination of knowledge from working many cases over the years, reviewing crime scene evidence and reports, studying the different types of offender, and becoming increasingly familiar with their motives and behavior.

The work is hard and involves long hours. But it is rewarding to contribute to a successful investigation and see justice served, or to help a family gain some closure in the case of a loved one.

Defense lawyer Melvin M. Belli, psychiatrist Manfred Guttmacher, and Roy Schafter during Jack Ruby's trial. Ruby was found guilty of murdering Lee Harvey Oswald and sentenced to death. He appealed but died in prison before the appeal could be heard.

If attention seeking had been the primary motive, it is probable that the bomb would have gone off in the bank, where it would have had the greatest destructive capacity and highest visibility.

Psychological gain, however, does not necessarily equate with unconscious or even secondary motive. Consider the following example:

> "Benny and Sheila were having trouble in their relationship. Benny was jealous and controlling; Sheila was pulling away. Benny engineered the theft of Sheila's jewelry from her home. In fear, she reached out to him, as he knew she would. Benny left the jewelry hidden at his office, rather than selling it."

In the previous example, Benny feared abandonment and the loss of his hold on Sheila. While he perpetrated the robbery, he made no effort to profit from the theft. He satisfied his goal by inspiring in Sheila a sense of need for him. This, the primary motive, is a psychological gain. Benny would not have robbed a bank, unless it would have been more likely to bring Sheila back to him.

Predatory Motives

Predatory criminal motives share the common feature of exploiting a vulnerable person or target for material gain. Profit is one of the most obvious of predatory motives. In larceny or robbery, the criminal gains money—and no matter how it is rationalized, when one person gains, others lose. Profit may also be a secondary motive, even to a psychological motive. In the previous example, if Benny had sold the jewelry, profit would have been a secondary motive. It would not have been primary because his goal was to protect his standing in his romantic relationship.

Proprietary gain is another predatory motive. Violent sexual assaults, particularly on strangers, clearly demonstrate proprietary sexual motivation. The criminal uses violence as a means of exerting power, control, and possession of the victim. Stalkers are proprietary, although their agenda is to further a pathological attachment, a romantic if perverse expression. Even if nonviolent, their actions are geared to insinuate themselves into the victim's thinking to control and perpetuate as intense a connection as possible.

Proprietary crimes encompass the range of material and psychological benefits that one person might want from another.

Psychiatrist Jean-Marie Abgrall gives evidence concerning the Order of the Solar Temple sect during the trial of Michel Tabachnik in 2001. Tabachnik, an internationally renowned Swiss musician and conductor, was arrested as a leader of the cult, and was indicted for "participation in a criminal organization," which included murder based on the accusation that he ordered a mass suicide.

> "Jeff Jennings was running for Senate. He faced a tough challenge from David Simon. Jennings engineered a break-in at the Simon headquarters to obtain his rival's list of potential donors. Jennings used this information to focus his negative attacks on Simon, and to compete with him at the heart of his fund-raising apparatus.
>
> With this new advantage, Jennings gained additional unanticipated donors and subsequently Simon was underfinanced. Jennings went on to win the Senate seat."

Power, property, collectibles, and prestige are all sufficient lures to motivate predatory crime.

Not all predatory crimes, however, are carried out for material or physical gain. For example, some sex offenders, particularly drug facilitated sex assaulters and child molesters, seek sexual gratification in their predatory acts.

Occasionally, there are reports of senseless crimes that are nevertheless predatory.

> "Jack wanted to feel what it was like to kill someone. So, one night, he drank a few beers and went in search of a victim. He resolved to kill a homeless person. Happening upon a drunken panhandler, he lured him to an alley, strangled him, and beat him to death."

In the previous example, the victim had no money and no property. There was no relationship between the criminal and victim. There was no sign of sexual gratification. Yet the criminal did seek out a victim to satisfy a curiosity, a thrill. This predatory crime was carried out for psychological gain, but with no proprietary motive.

Attention-seeking crimes may also be predatory in nature. In such cases, criminals typically want to promote political causes. Popular movies suggest that there are killers who search out victims to play games with police. However, seeking attention without an apparent agenda or another motive is rare. Those guilty of such crimes are usually disorganized and do not devote as much effort to escaping police as they do to the spectacle of their crimes.

> "James hated whites. He plotted that one day he would shoot as many white people as he could, before being killed himself. Over a period of months, he wrote a list of his gripes with the whites in his neighborhood. When he finally embarked on a shooting rampage, he left the list behind, knowing it would be found and that there would be major public attention devoted to scrutinizing the reasons for his action. However, James was not killed. Instead, he was cornered by police, who refused to engage him in a gun battle. Eventually, they subdued him after he ran out of bullets and grew fatigued from the excitement of the drama."

Reactive Motives

A response to a provocation or perceived provocation is a common motive. The likelihood of a reactive motive warrants more consideration when financial or material gain is unclear.

Vigilantism is a vivid example of reactive motivation, driven by revenge and retaliation. More common, however, are responses to slights or actual attacks. Reactive crimes often occur during periods of inflamed emotion, uninhibited drug use, or emotional vulnerability. For this reason, they are often impulsive and manifest more disorganized modus operandi. That said, reactive crimes, particularly when carried out by those who frequently break the law, may follow considerable planning.

> “Gary was out drinking one night and bumped into Peter at the bar. Peter humiliated Gary by throwing him to the floor, and walking away. Gary left the establishment and hid some distance from the entrance with a baseball bat. When Peter came out, Gary crept up behind him and struck him down.”

The motive was reactive; the crime was not impulsive. In this instance, Gary had no issue with Peter prior to their confrontation. His humiliation inspired the volcanic reaction, although he obtained a weapon and waited patiently to stalk and then attack Peter in retaliation.

Likewise, poorly planned and impulsive crimes may nevertheless be driven by predatory motives. Opportunistic offenders, those who devise a modus operandi to exploit circumstances they encounter, may be impulsive in their decision making.

Crime scene warning tape circles a chalk body outline at a crime scene.

A Day in the Life: Forensic Psychologist

Dr. Katherine Ramsland, Assistant Professor of Psychology, Pennsylvania

A man stabs his wife to death and claims he was sleepwalking. A girl says she was held captive for years inside a box. A man confesses to crimes he did not commit. A mother drowns her children to rid the world of Satan. These are typical of the cases I encounter as a forensic psychologist, and I may be called in to provide an assessment of the case for the prosecution or defense, to testify in court or to treat the criminal.

Working as a forensic pyschologist can be intriguing, and no two days are the same. While most forensic psychologists are clinicians who specialize in forensic matters, the discipline actually involves a range of specialties in civil and criminal matters. These include consulting on criminal investigations, assessing threats of violence in schools or workplaces, determining the fitness of a parent for guardianship, developing specialized knowledge of crimes and motives, evaluating the effects of sexual harassment, and conducting forensic research.

Some days, I have to appear in court, where I am often asked to evaluate a person's psychological state to determine if he or she can participate in the legal process, or his or her mental state at the time the offense was committed. Another common task is to appraise a defendant's behavior, such as malingering, confessing or acting suicidal. As a consultant, I may also help a lawyer select jury members.

It is important to understand the workings and expectations of the court. As an expert witness, I must be credible, confident, competent, and prepared. Courts prefer clear decisions, jargon-free assessments, and objective information that directly addresses the issues concerned.

There is a popular assumption that forensic psychologists track down serial killers, but this idea comes from fiction, not real life. Although psychologists may interview serial killers for research or preparation for a court case, they are not detectives, and most are not profilers. Nevertheless, the FBI's Behavioral Analysis Unit does provide training in criminal psychology to agents who will act as crime-scene consultants.

A good profile is an educated attempt to provide investigators with a clear impression of the type of person who committed a certain crime or series of crimes, based on the idea that people act within their unique psychology and inevitably leave clues. From a crime scene, a profiler can assess whether the offender is an organized predator or someone who has committed an impulsive crime of opportunity. He or she may also determine if the offender used a vehicle, displayed criminal sophistication or was addicted to a sexual fantasy.

Profiles work best if the offender displays obvious psychopathology, such as sadistic torture or mutilation of the body after death. Some killers leave a "signature"—like always posing a corpse or tying a ligature with a complicated knot. This helps to link crime scenes and alert law enforcement agencies to the presence of a serial criminal.

Motiveless Crimes

Is it possible for a crime to occur without a motive? Perhaps this is the embodiment of the senseless crime. The term senseless is typically used to convey an idea that a crime had no purpose. More often than not, however, it means that the crime accomplished nothing. The criminal may regret the crime after the fact, but at the time may have had specific short-term objectives that inspired the criminal choice. The notion of a motiveless crime evokes immediate consideration of psychiatric issues. If there is no motive for a crime, a person's behavior is at least irrational, and consciousness may be severely impaired.

Psychosis and Motive

Psychosis is a manifestation of irrational ideas, perceptions, emotions, communication, orientation, or behavior. All have a bearing on criminal motive and the expression of it. Psychotic ideas are known as delusions, and a delusional individual maintains specific ideas despite all evidence to the contrary. The content may be persecutory, jealous, romantic

Case file:

Arthur and Jackie Seale

On April 29, 1992, Sidney Reso, president of Exxon International, was kidnapped as he left his home in New Jersey. A ransom note instructed the company to arrange a cell phone account to receive further instructions. The kidnappers claimed to be environmental activists with a grudge against Exxon.

They had made a mistake, however, because calls to cell phones can be traced. A surveillance operation was mounted, concentrating on local pay phones. A man and woman made calls to the cell phone and delivered more notes, demanding $18.5 million, but the police and FBI were unable to catch them.

(erotomanic), or grandiose. Crime driven by a delusion of persecution is typically reactive; delusional erotomanic crimes are proprietary, and crime driven by grandiosity may compel a variety of motives.

Deeper probing, beyond the mere existence of a delusion, can resolve why a belief drove a person to crime. The reason that person made a criminal choice may not necessarily be explained by the delusion itself, unless the person was convinced that he or she was involved in a criminal enterprise, perhaps as a mobster or terrorist. Consider the following:

> “Lisa, a dentist, believed that a former client was following her. While there was no evidence of this harassment, Lisa's story elaborated that the client, Marcia, had bugged her phone, placed cameras in her living room, and was tampering with her feminine hygiene products. One day, Marcia was found shot to death; Lisa admitted that she was the killer.”

An FBI profiler was called in to study the evidence and concluded from the phraseology of the notes that the writer had a security background and some connection to Reso or Exxon. The way in which Reso had been snatched suggested careful planning, but the reliance on a cell phone indicated inexperience in this kind of crime. Environmentalists were discounted, and the case was seen as straightforward extortion.

Delivery of the ransom was agreed for June 18, with the FBI and police keeping watch. A man and woman were seen using pay phones that were identified as the sources of calls to the cell phone. One of them was spotted driving a rental car, but then was lost. Eventually, they were arrested when they attempted to return the car to the rental office.

The suspects were Arthur and Jackie Seale. Arthur, a former policeman, had worked for Exxon as a security officer and had seen kidnapping Reso as an easy way of obtaining enough money to pay off substantial business debts. A week later, Jackie led police to where Reso was buried. The couple had never intended to murder him, but he had died from a heart attack as a result of their treatment.

Arthur Seale pleaded guilty to murder, kidnapping, and extortion, receiving a life sentence; his wife Jackie was given 20 years for extortion.

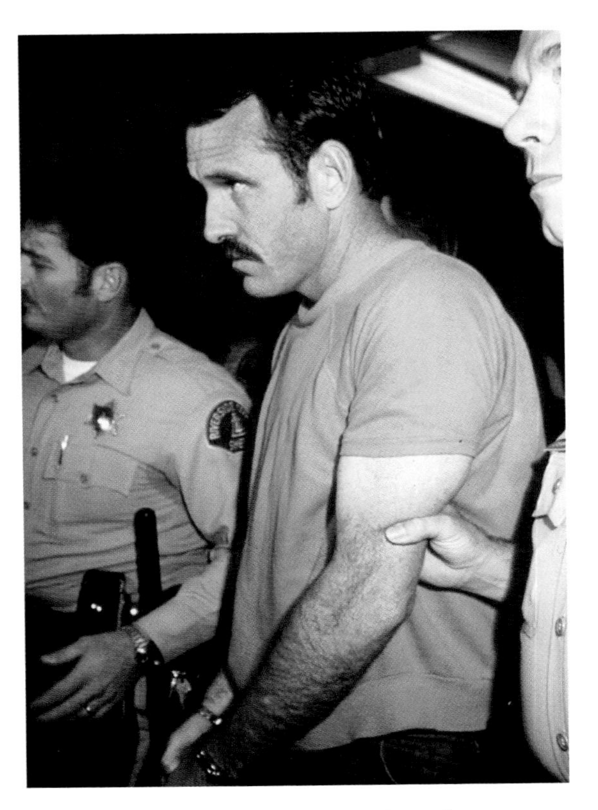

David D. Hill is arrested for the "Trash Bag Murders" in California. Hill and his lover, Patrick Kearney, lived in a meticulously clean bachelor pad from where they planned homicidal escapades from 1975–77. When the couple saw a wanted poster of themselves, they walked into the Sheriff's office and confessed. Hill was released for lack of evidence. Kearney shouldered the guilt and confessed that killing "excited him and gave him a feeling of dominance."

Why did Lisa shoot Marcia when she did? Did she feel threatened? Was it fear of being killed? Was it a revelation that Marcia was part of something bigger than the two of them? Was it out of anger?

Many individuals with major psychiatric illnesses develop significant hostility. During times when the effects of the illness are more pronounced, the person may act on that anger, with no regard for the consequences. The provocation may be substantial or minor. Some who suffer delusions also think they hear voices, which may encourage or order them to break the law. Senseless crime? Sure, when you consider that the drive to act stems from the orders of something that does not exist. Is the crime motiveless, however?

Without examination, the motive remains unclear. Even when hallucinations are of the command type, the people who experience them do not necessarily follow their directives without critical thought. Assessing the motive for following the hallucination's instructions at a particular time resolves the motive itself.

Psychosis expressed as irrational emotion may also act as a motive. Some who are psychotic may produce an unjustified extreme reaction to relatively minor provocation. This scenario is more frequently associated with intoxication or substance abuse.

> "Manuel was jilted by Marisol, his girlfriend. He went to her apartment building and rang her bell incessantly to be let upstairs so that they could talk. She ignored him. A neighbor called out to him to stop. Enraged by the neighbor, Manuel retrieved a gas can, returned to the apartment building, and set it on fire."

Psychotic crimes, therefore, are not necessarily motiveless at all. One situation that does pose that possibility, however, results from the disorientation and confusion of delirium. While suffering acute medical conditions, some individuals may develop delirium, an altered state of consciousness that muddles rational thought. Acutely confused individuals may carry out purposeless crimes, such as random physical blows or grabbing.

Resolving Questions of Motive

The examiner must keep an open mind and look beyond seemingly obvious motives; the criminal's admitted motive may merely be the easiest for him or her to reveal. Closer probing can uncover other motives that the person would rather conceal because they may be psychologically or legally harmful.

Reviewing the criminal's history acquaints the examiner with any conflicts, sensitivities, or ongoing or building issues. This information is relevant when determining the motive for the offense. The criminal reaches the event in question as a product of past experience and motivations. Not even the wisest examiner can divine motivation without untangling the criminal's life, as seen in the example below.

> “Craig had been married to Wendy for 20 years. He had no criminal record and worked as an engineer in a Bible-Belt town. The couple had financial problems; with four children, they had struggled to keep up appearances. Although their marriage was seen as devoted, Craig had embarked on a plot to kill his wife and family, set the home on fire, and burn it to the ground. He succeeded in murdering his wife, and then attempted to kill his children, but tired before he could do so. Interviewed later, Craig said that he had been ashamed and wanted to save the family from poverty.
>
> The defense worked hard to limit the probing of Craig's background, citing its irrelevance. Continued close exploration, however, revealed that he had lived a double life as a homosexual for 30 years, without the knowledge of his wife and children. His relief, in the aftermath of the destruction of his family life, was that of a man who had escaped, and he spoke openly of the difficulty of living a secret life.”

Criminal Profiling

Recent books, television shows, and movies have brought a new crime-solving hero into the limelight—the profiler, who is able to lead detectives to violent serial criminals by careful observation and deduction. While it is true that profilers can play a very useful part in solving serial crimes, in reality this does not happen as often as the fictional accounts would have us believe. Most serial cases are solved by good old-fashioned detective work and forensic investigation.

A criminal profiler will usually be called in on a case when traditional investigation methods are not making headway, perhaps because there is little hard evidence to point to a potential suspect. His or her purpose is to steer the investigation in the right direction, helping to narrow the search for a suspect. The profiler does this by studying all the evidence from the crime scene and coming to conclusions about the type of person who committed the crime. These can include the criminal's background, intelligence, education, age, and possibly even job and marital status.

Many of these deductions are based on statistical information gathered from past crimes, the assumption being that certain types of crime, committed in the same manner, are carried out by people with similar personalities. A profiler's knowledge of criminal psychology and his or her experience of past cases also play a major part in the preparation of a profile. Profilers are often very effective in the investigation of serial crimes, simply because these crimes produce more information to work with.

Violent criminals often leave more information to indicate their personality at crime scenes than lesser offenders, because their violent crimes are often personality driven. A series of frenzied attacks on women, for example, could indicate that the killer had a deep hatred of women, prompted by some experience in his past, such as a girl's rejection as a teenager, or perhaps having a domineering mother.

The study of serial crimes by a profiler can lead to the identity of signature or common features at each crime scene. These are useful in attributing a series of crimes to a particular offender, and can lead to the inclusion of crimes beyond the initial scope of an investigation.

A Day in the Life: Geographic Profiler

Dr. Kim Rossmo, Professor and Geographic Profiler, Vancouver, Canada

It is the first day of my visit to a city I have never seen before. By the time I leave, I will have visited places most lifelong residents won't even be aware of—nor want to. We will be working until the early hours of the morning, so the detectives and I meet at noon, coffee cups in hand. We gather around a conference table in the Task Force Headquarters. There are only a few geographic profilers in the world, all from major police agencies, so we often travel to far-flung places to provide assistance in cases of serial murder, rape, bombing, and arson.

Because geographic profiling is relatively new, I start with a presentation on its theory and methodology, capabilities, and limitations. Then it's time to hit the streets. We visit the locations connected to a serial murder case, viewing the victim encounter and body dump sites. My goal is to answer questions related to how the offender hunted for his victims, why he knew these areas, what is his comfort zone, and—most critically—where he lives.

Geographic profiling analyzes crime locations to determine the most likely area of offender residence. It cannot solve a crime (only physical evidence, a witness, or a confession can do that), but it can help investigators manage information and prioritize suspects. Other techniques, such as DNA testing, can then be more efficiently applied. Because geographic profiling is an investigative support technique, we rarely go to court. Underlying this work is an extensive body of research on human behavior and the geography of crime. So it is not surprising that we often work closely with psychological profilers. They provide the "who," and we provide the "where."

During the visit to the crime sites, I look for geographic clues: street routes, arterial roads, jogging paths, transit lines, late-night bars. As the murders in this case happened at night, our visit will be repeated in the dark. The attacks of even the most bizarre serial killer are not random. Rather, there is an underlying pattern to how the killer hunted for his or her victims and where he or she chose to commit the crimes. By decoding that pattern, we can help investigators find who they are looking for.

My visit will conclude with the preparation of a preliminary geoprofile and a brainstorming session on possible investigative strategies. Later I will prepare a written report.

The major tool of the geographic profiler is a software system called Rigel. This program performs the millions of calculations of the criminal hunting algorithm necessary to produce a geoprofile. We input the crime locations, and Rigel outputs maps, with areas of offender residence likelihood marked by different colors. Other tools of the trade include a global positioning system (GPS) to determine the locations of remote body dump sites, a digital camera to visually record the surrounding areas, a laptop computer, hiking boots, and a cup of good coffee.

07 The Arrest

Sometimes the person who is responsible for a crime may be obvious, but more often a lot of diligent and exhaustive detective work is required before forensic investigators can identify one or more suspects and arrest them.

The criminal justice system and the process by which it is driven are based on a series of important decisions that are critical to both the individual concerned—the accused—and the society that the system is designed to protect. These decisions follow a neat pattern of evidentiary burdens that are observed at every step of the criminal process.

Before looking into the various phases of this system, it is important to understand these basic evidentiary concepts and how they impact upon a police investigator's decision to pursue a criminal homicide charge against a person, as well as the homicide prosecutor's decision to fully and fairly prosecute a defendant to conviction. The concepts are common to the legal systems in the United States and Canada.

The first requirement is to establish the *corpus delicti* of the crime, or translated from the Latin, the body of the crime. As a rule, a two-part test is applied when attempting to determine *corpus delicti*. First, it must be shown that a crime has been committed by someone; second, that someone undertook the crime with the requisite mental fault, the *mens rea*, that resulted in a specific injury or loss.

Most *corpus delicti* problems arise in homicide cases because of two specific circumstances. One is the absence of a body in a purported homicide based upon circumstantial

A handcuffed and arrested man is taken into custody.

evidence. In this instance, it is very difficult to be absolutely certain that some day the victim will not show up alive and well. For this reason, in America, many states have rules that prevent the imposition of capital punishment without direct proof of the *corpus delicti*. Alternatively, the body is present, but the examination of it and the surrounding circumstances fail to establish whether the death was caused by criminal behavior or by some other means.

The *corpus delicti* rules are widely applied and generally observed in all jurisdictions for one simple reason: to reduce the possibility of punishing a person for a crime that, in fact, was never committed.

The next critical concept concerns the burden of proof. In the early stages of a criminal investigation, police officers and other state-related investigators are permitted to stop and question, or stop and frisk, certain individuals only if it can be shown that there are reasonable grounds to suspect criminal activity. In serious cases such as homicide, the decision to charge a suspect, and issue a warrant for his or her arrest and incarceration, is based on the burden of showing probable cause. Is there sufficient evidence to indicate that a crime has been committed, and that the accused is the most likely person to have carried it out?

The greatest of all evidentiary burdens is proof beyond a reasonable doubt in court that a suspect murdered a victim. The homicide prosecutor must decide whether all the facts and evidence would lend themselves to the ultimate conviction of a criminal defendant. That decision must be made with the knowledge that the evidence will be scrutinized by a judge or a jury, who need to reach the same conclusion to return a verdict that results in conviction.

Pre-Arrest Procedures

Once police investigators are made aware of a crime and identify a possible suspect, they are responsible for making the initial determination of probable cause, which forms the basis for the subsequent arrest of the killer. The probable cause standard is a twofold test and lays down the baseline for all future criminal proceedings. The investigators must determine:

1. Whether a crime was actually committed.
2. If a crime was committed, whether there is sufficient

Case file: John Prante

In June 1978, Mark Fair and a friend arrived at the new home he and his fiancée, Karla Brown, had bought in Wood River, Illinois. The couple had only moved into the house on the previous day, and Karla had taken the day off work to arrange their belongings. To their horror, the two men found Karla in an upstairs room, almost naked, with her hands tied behind her back and her head thrust into a drum of water. She had been strangled and struck on the head with a hammer.

The police were called, and they found a large number of fingerprints in the house, but the couple had thrown a housewarming party the night before, and there was no way of knowing which, if any, might have belonged to the murderer. Among the suspects questioned were Paul Main, the couple's new neighbor, and his friend, John Prante, both of whom had watched the couple moving their things into the house. They were given lie detector tests; Prante passed his, but Main's was inconclusive.

Subsequent computer enhancement of photographs of Karla's body showed the presence of bite marks on her neck. The police felt that Main was the strongest suspect because of his inconclusive lie detector test, and he was asked to provide a bite impression, but this proved that he was not the murderer.

evidence to identify the alleged killer and justify arresting and charging that person.

Was a Crime Committed?

Previous chapters in this book have dealt with the responsibilities of the first responders to the death scene, the homicide detectives, the paramedics or emergency medical technicians, and the death investigators. It is from the combined efforts of these specialists that the coroner or medical examiner will make the critical decision that may, or may not, lead to criminal proceedings—the manner of death.

The manner of death is a legal conclusion based upon the facts and circumstances surrounding the death. To recap, in almost every medicolegal jurisdiction in the world, there exist only five manners of death: homicide, suicide, accidental, natural, and, in cases where insufficient evidence exists to reach a conclusion, undetermined.

Consider an actual case that occurred in Pittsburgh, Pennsylvania, under the jurisdiction of the Allegheny County Coroner's Office.

Little further progress was made in the case until 1982, when an FBI profiler was called in. Among his conclusions were that the crime scene had been staged to look like a sex game that had gone wrong, and that Karla had been murdered in a fit of rage because she had rejected her attacker's advances.

The police mounted a media campaign in an attempt to force the killer into the open. Newspapers reported that a profiler was involved and that the police were going to exhume the body to recover additional evidence. The cemetery was watched for anyone acting strangely. After the exhumation, the police told the press that the body was in excellent condition. All this was expected to put pressure on the killer, affecting his behavior, and it was hoped that this would be noticed by someone.

The publicity campaign paid off, leading the police to a witness who had met a man who claimed to have been in Karla's house on the day of the murder and had seen a bite mark on her shoulder. That information had never been released by the police. The man was John Prante, who had been behaving as if under extreme pressure for some time. He was arrested and compelled to provide a bite impression, which matched the mark on Karla's shoulder. He was sentenced to 75 years for murder and burglary with intent to rape.

> "A young man was discovered lying in a driveway near an apartment building just off the campus of the University of Pittsburgh. He had suffered a fatal gunshot wound to the head, and two semiautomatic pistols were found at the scene. One pistol was no more than 2 ft. (60 cm) from the body; the other was about 10 ft. (3 m) away. Three people of similar age to the deceased were seen fleeing from the scene.
>
> Response to the scene was immediate: police officers and detectives happened to be within blocks of the incident, and a deputy coroner was also nearby."

What direction would this investigation take? It is accepted practice that these scenes are approached initially as a homicide, but other factors and considerations come into play when determining the resources to be allocated to the immediate investigation. This scenario suggested the possibility of at least three of the five manners of death: homicide, suicide, and accidental.

The scene that greeted the initial investigators gave a first impression that the death was a homicide. Two weapons were present, and one was a short distance from the body. Could the deceased have been the victim of mutual combat and, if so, was he or the killer the aggressor? Did the killer drop his or her weapon at the scene? Was the flight of the three individuals from the scene an indication of their guilt?

On the other hand, the incident could have been an accident. The young man might have been carrying the pistols from his apartment to his car and tripped on the driveway curb, causing the accidental discharge of one of the weapons. Or, as has happened in a number of reported cases, he may have been feigning a suicide attempt or playing a practical joke for the benefit of his companions, unaware that the weapon was loaded.

Finally, it could have been a suicide. Upon initial examination, the head wound appeared to be of close contact, and one of the two weapons was very close to the body. In addition, male suicide is often carried out through a violent and public display. Fortunately, the early arrival of a deputy coroner allowed for an immediate examination of the body and the contents of the young man's pockets by the police. In addition to a number of live cartridges, the examination revealed a driver's license that listed the adjacent apartment building as the young man's address and a key to the apartment. Detectives found a suicide note in the apartment, and discovered that he had packed all his belongings and left instructions for their disposal. Later, the detectives discovered that the young people seen fleeing the death scene were students who had been walking home from the university, and who had been unwitting and terrified observers of the young man's tragic final act.

This example illustrates the need for cooperation between law enforcement and forensic investigative agencies at this critical point in the criminal justice process. Once the manner of death is established, the criminal aspects of the investigation may either come to an end or move forward. In this case, the manner of death was ruled as suicide. However, if the investigators had been confronted with a homicide, a criminal act, they would have been compelled to address the second part of the probable cause test: assembling sufficient evidence to indicate the guilt of a suspect, allowing that person to be arrested and charged with the crime.

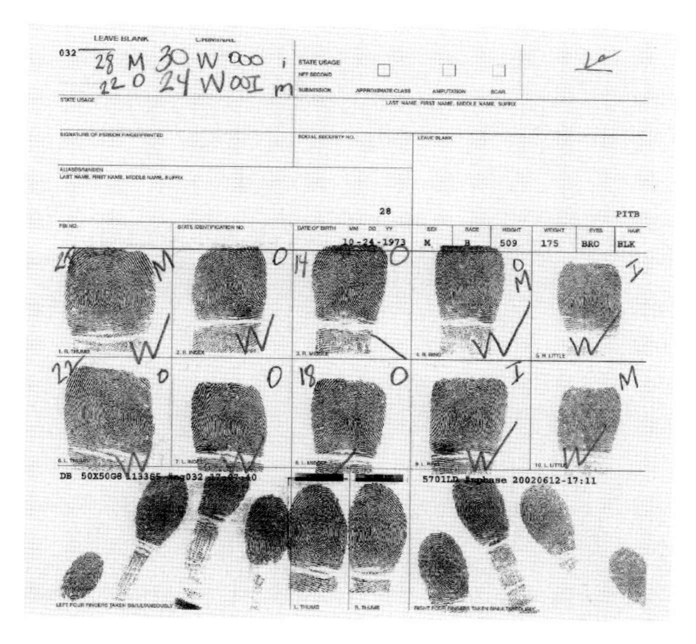

LEAVE BLANK

032

STATE USAGE

28

PITB

DATE OF BIRTH 10-24-1973

M B 509 175 BRO BLK

When someone is arrested, a fingerprint card is created and kept on file. The first systematic use of fingerprinting to aid in criminal investigation was in 1902 by the New York Civil Service Commission.

Establishing Probable Cause

Some crimes actually occur in the presence of police officers, allowing for an on view arrest, without the need for a warrant. In such cases, the investigation involves no more than recording the officer's observations of the crime. This clearly establishes probable cause. However, criminal homicides are usually more complicated and detectives must use a variety of pre-arrest investigative techniques to develop sufficient information to complete the probable cause test.

Every seasoned homicide detective knows that a case can be made or broken in the first few hours following the commission of a murder. If the investigation stretches into days or even weeks, evidence may be spoiled and potential witnesses may disappear or become reluctant to participate.

In homicide cases it is crucial that trace evidence such as fingerprints, blood, and DNA is promptly collected and properly preserved. The analysis of a single fingerprint, hair, or DNA sample from a bloodstain has been deemed sufficient evidence to establish the requisite probable cause to link a person to a crime.

In recent homicide cases, eyewitness testimony has proved to be the least credible evidence in establishing the guilt of a suspect. Also, police informants may be equally unreliable. When using an informant, the investigating officer must carefully consider the past criminal conduct of that informant as well as any ulterior motives he or she may have to provide false information. At this stage, however, the detective is only trying to establish probable cause for an arrest—not necessarily the suspect's culpability beyond a reasonable doubt. It is essential to identify and interview eyewitnesses, or develop police informants as quickly as possible following the discovery of the crime.

Police and forensic officers guard a house in Manchester, United Kingdom where a police officer was fatally stabbed during the arrest of three terror suspects in January 2003.

A Day in the Life: Crime Reporter

Paul Anderson, Police Reporter, *Herald Sun*, Melbourne, Australia

In many ways, being a crime reporter is like working as a detective. When a crime occurs, the key is to get to the location as quickly as possible and canvass the scene. Just as police investigators try to determine the events as they unfolded in the hope of solving the crime, I attempt to piece together the story to provide the most comprehensive coverage possible. I ask neighbors and eyewitnesses to describe what they have seen and heard, and to give information about anyone involved—both victims and offenders. It's all about keeping my eyes and ears open.

The lifeblood of any crime reporter is his or her contacts, preferably on both sides of the law. While a police investigator will brief the media at a crime scene, a good reporter should be able to discover more details of the case from contacts and not simply rely on officially released information.

I usually hear through my contacts and sources about incidents as they are happening. Another source of information are police radio scanners, which are always crackling and babbling in our office. We hear police in the field notifying the communications center of their movements and crime scenes that they are attending. It is important to know the police call signs and their most frequently used incident codes to be able to keep abreast of events. The police media liaison unit is also helpful in keeping us aware of what is happening in the world of crime. If I'm not in the office, I'm usually hanging around a police station or having a beer with a contact.

Scouting around a crime scene for the story, and asking questions of anyone and everyone, is the most basic element of my work. When not racing to a scene, I follow up on previous stories and keep abreast of unsolved investigations. Just like a detective who responds to a crime, I stick with a case and continue to write about it until it is solved. The better rapport a reporter has with his or her police contacts, the more inside information he or she will receive. It can take years to build the trust necessary to form such a relationship.

For some crime reporters, the hardest aspect of the job are what we call "intrudes." These are stories that require us to knock on the doors of people who have lost loved ones through horrible crimes or tragic accidents. Intrudes are tough, but the stories that come out of them are designed to stir public interest and help catch the offenders.

As a crime reporter, I have to know the law, particularly as it relates to defamation and contempt of court. Once a case enters the court system, the court reporters take over. If the accused is a notorious criminal, however, or has committed a particularly callous or vicious crime, I prepare feature stories on him or her to run once the trial and any appeals are over.

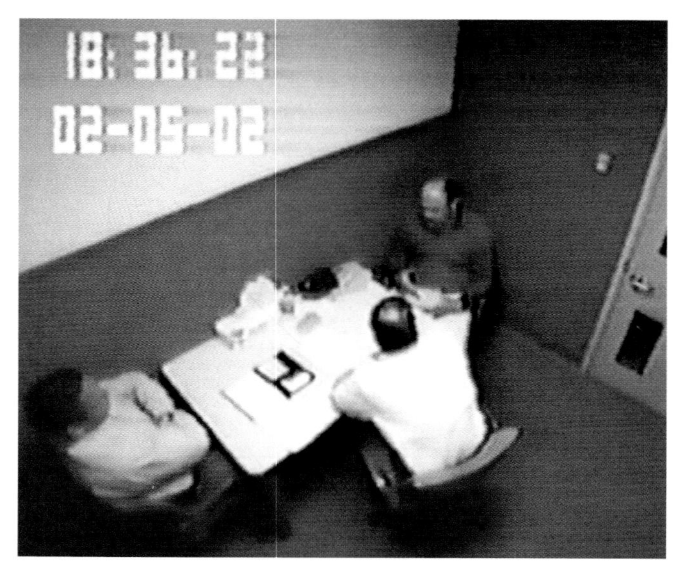

Two police officers interrogate David Westerfield in 2002 in California. Westerfield was questioned in the disappearance of Danielle van Dam, one of his neighbor's children. Danielle was later found murdered and Westerfield was charged, convicted, and sentenced to death for her murder and kidnapping.

Buying time

While, in the long run, such testimonial evidence may fail to establish the requisite proof for conviction, it is sufficient legally to establish the probable cause for an arrest. This allows for further post-arrest investigation, which may involve time-consuming techniques and procedures that will firmly establish the proof beyond a reasonable doubt. In many circumstances, time is required to obtain judicial authorization to conduct thorough searches of a suspect's home or place of business. Evidence collected at a crime scene may require sophisticated testing to bolster the prosecution's efforts to meet the standard of beyond a reasonable doubt required for a conviction.

The ability to establish probable cause for arrest provides a very important investigative source that normally is unavailable beforehand—the arrested person. The suspect may be compelled or persuaded to offer additional incriminating statements beyond those assembled to establish probable cause, or the suspect may give valuable information that leads to the identities and subsequent arrest of any co-conspirators.

The suspect may be placed in a lineup in which he or she may be identified by witnesses, or the victim in cases of rape or assault. Alternatively, witnesses may simply view the suspect on their own in what is known as a show-up. Also, the suspect may be compelled to provide DNA or hair samples for comparison with evidence collected by the forensic investigators at the scene of the crime.

The Interview and Interrogation of Suspects

The confrontation between a police interrogator and a suspect is fraught with problems. A mishandled police interrogation—either in legal terms or in its practical application—can result in a tainted confession or one that is excluded at the time of trial.

It is important to distinguish between an interview and an interrogation. The players in each process are the same—the police investigator who asks the questions, and the suspect or witness who supplies the answers, which can provide either crucial information or a confession of guilt. Considerable skill and subtlety are required on the part of the investigator if an interview is to move on to the interrogation phase and achieve the desired result.

An interview is a structured, non-confrontational process. Fault, accusation, and culpability are not at issue. The skilled interviewer addresses the suspect with specific questions that are designed to elicit behavior patterns and certain answers that are consistent with guilt or innocence. This establishes the so-called emotional parameters that subsequently will control the course of the interrogation. It helps the interviewer to determine which emotional buttons to push and which ones not to push. During the interview, the investigator may develop fact patterns concerning the case and glean information that could incriminate others or help locate more witnesses.

Moving into the interrogation

Once the initial interview has been completed, the investigator has established a rapport with the suspect, and will be able to analyze and recognize certain physical and emotional behavior patterns. At this stage, the interview shifts into the interrogation phase.

When confessions are barred

A confession may be barred from admission as substantive evidence in a criminal proceeding for three reasons:

1. The methods used by the police to extract the confession (physical violence, threats, torture, intimidation) were so dubious that the confession is highly unreliable.
2. The methods used by the police were offensive in and of themselves, even though the reliability of the confession is not questioned.
3. The person offering the confession was seriously impaired, either by mental infirmity or induced incapacity, regardless of the lack of wrongdoing by the police or the trustworthiness of the confession.

In comparison to the structured nature of the interview, the criminal interrogation has a decidedly different dynamic. The interrogator adopts a confrontational tenor, becoming accusatory toward the suspect. From that point, the forensic investigator's only goal is the suspect's confession and acknowledgment of guilt.

The purpose of both the interview and interrogation is to identify the guilty party and to obtain a legally admissible confession. When faced with an uncooperative suspect, the skilled interrogator can usually determine, through apparently innocent answers, whether a suspect is likely to have committed the crime. This is important in focusing the investigation when faced with a large number of potential suspects.

Police interrogations are tricky. The only way a case can be solved by interrogation is through a legal confession by the guilty party. This is especially important in cases where police and forensic investigations have failed to provide a single piece of evidence. Criminals rarely admit their guilt unless subjected to long periods of interrogation under the strictest

Case file:
Darrell Devier

In December 1979, 12-year-old Mary Frances Stoner was abducted from outside her home in the small town of Adairsville, Georgia, after getting off the school bus. Some days later, her body was found in secluded woodland about 10 miles away. She had been raped, strangled, and finally killed by several blows to her head with a rock, which was lying nearby. Although she was fully dressed, her clothes were in disarray, suggesting that she had been made to undress, assaulted, and then allowed to put her clothes back on.

Initial investigations proved fruitless, so the homicide detectives called in an FBI profiler, John Douglas, in the hope of learning more about the type of person who had committed the crime. The profiler's analysis of the evidence concluded, among other things, that the crime had been opportunistic and carried out by someone who lived in the area—the place where the body had been left suggested local knowledge—who had probably killed the girl to prevent her from identifying him.

When they considered the entire profile, the detectives realized that it was a very close match to someone they had already interviewed as a possible witness to the crime, Darrell Devier. He had been working close to the girl's home on the day of her abduction, trimming trees. Moreover, some time before, he had been

conditions of privacy. It is against human nature to incriminate yourself. Unsolicited and spontaneous confessions may happen from time to time, but they are certainly not the norm.

The interrogator, however frustrated by an evasive or uncooperative suspect, must always be mindful of the legal constraints of the confession process. In the United States, prior to any custodial interrogation, the interrogator must explain the suspect's rights through the Miranda warning. In other countries, such as Canada, suspects must be read their rights upon arrest. The suspect must actually be in custody for the interrogation, since the investigator's conduct during a noncustodial interview will differ greatly from that of a full-blown custodial interrogation. The interrogator must also bear in mind the exclusionary rule, or the fruit of the poisonous tree doctrine, which would exclude any confession from a criminal proceeding if it had been obtained in a manner that violated the suspect's rights.

What complicates the matter further is that there is no specific conduct that, by itself, renders a confession

accused of raping a 13-year-old girl, but the case had been dropped because of insufficient evidence.

Devier, however, exuded self-confidence, was evasive and his polygraph results were inconclusive. The investigators needed a means of rattling him to determine if he actually was the killer. They decided to put him at his ease by interviewing him at night in a softly lit room. To suggest that they had built a case against him already, they loaded the desk with stuffed folders bearing his name. The key item, however, was the bloodstained rock found next to the girl's body. This was placed so that Devier would have to turn to look at it, giving his interrogators an opportunity to look for signs of nervousness on the part of their suspect.

The ploy worked. When Devier noticed the rock, he began breathing heavily and sweating profusely. His interrogators let him believe that they had evidence of bloodstains on his clothing that tied him to the crime and that they knew he had not meant to kill Mary Frances. He became defensive, and under further close questioning eventually confessed to the rape and killing of Mary Frances, as well as to another rape charge the previous year.

Devier was convicted of the girl's murder and finally executed in 1995.

inadmissible. The courts have a long-standing history of looking at the totality of the circumstances in confession cases, including the characteristics of the accused and the details of the interrogation. It is the obligation of the police interrogator to be aware of these parameters and to apply them accordingly. What may prove to be a small victory at the beginning of the investigative process can become fatal to the ultimate prosecution of the case.

Filing the Criminal Complaint and Charging the Suspect

Once it has been established that a crime has been committed and sufficient information has been gathered about the identity and culpability of the person accused of the crime, the homicide investigator sets out the facts of the case in an affidavit of probable cause, which is submitted to the lowest level court—generally the magistrate court. There, the magistrate, or the issuing authority, conducts an *ex parte* (one party) review of the case. This is done for the benefit of one party only (the investigating officer) to the exclusion of any other interested party (the accused).

The purpose of the review is to ensure that the affidavit provides sufficient incriminating information to establish probable cause for the arrest warrant to be issued, and the suspect to be taken into custody.

In many cases in the United States, when an investigating officer is unable to compile sufficient information or evidence to support an affidavit of probable cause, a suspect may be charged following an indictment by a grand jury proceeding. As a rule, the grand jury is a panel of ordinary, private citizens who are selected to review certain cases. The grand jury may be assembled to review a single case, or it may sit for weeks or months and consider a number of cases.

While the grand jury process is similar to the magisterial *ex parte* review, it is based on totally different procedures. The grand jury meets in secret, and the evidence is presented by a special prosecutor. Because the proceedings are private and the government has the power to compel the appearance and testimony of witnesses, significantly more information and evidence are likely to be available to establish the probable cause for arrest than in a magistrate's court. The grand jury is a very effective and powerful tool in establishing a complicated case in the face of reluctant witnesses and an initial lack of credible evidence.

Once a suspect has been taken into custody and charged, he or she becomes the defendant in the criminal process and is immediately brought before the magistrate at the first appearance, or the preliminary arraignment. Normally, this is a very brief proceeding. The magistrate ensures that the person presented is the person named in the complaint, informs the defendant of the charges by reading the complaint, and explains the various rights that the defendant may have in future proceedings.

Usually, the magistrate advises the defendant again of his or her rights. The defendant has the right to remain silent, and anything said to the court at that time may be used against him or her at the time of trial. The defendant also has the right to counsel; if he or she cannot afford counsel, an attorney will be appointed to represent him or her. The magistrate may also inquire about the defendant's

The Greenwich Conneticut Police Chief briefing media after Michael Skakel turned himself in to police after an arrest warrant was issued in the 1975 murder of Martha Moxley.

financial standing and may make an immediate determination of need. In many cases, the magistrate will contact the Public Defender's Office or arrange for appointed counsel. In Canada, suspects have a right to access to counsel immediately upon arrest and before making any statement.

Finally, the magistrate tells the defendant the time and date set for the preliminary hearing and, if warranted, sets bail. As a rule in homicide cases, bail is unavailable to the defendant, and he or she is kept in custody until the preliminary hearing.

The Role of the Media

The advances in communication technologies, the proliferation of competitive news media outlets, and the sensationalism of high-profile murder cases in recent years have created the need for increased sensitivity to the role of the news media in homicide investigations. On one hand, the news media can be of great help to a homicide investigator; on the other, it can have a devastating effect on the course and ultimate outcome of an investigation.

Case file: Timothy McVeigh

On April 19, 1995, a truck packed with homemade explosives destroyed the Alfred P. Murrah Federal Building in Oklahoma City, killing 168 people, many of them children. At first, it was assumed that a foreign terrorist group had been responsible. But then the news media realized the significance of the date—it was the second anniversary of the FBI's storming of the Branch Davidian compound in Waco, Texas.

Among the vast amount of evidence recovered from the site was a piece of metal bearing a vehicle identification code. This had come from the truck used by the bombers, and it led investigators to the Ryder Truck Rental Office in Junction City, Kansas. The staff there provided descriptions of two men

In years gone by, criminal cases were generally left to a newspaper beat reporter, who spent his or her days or nights at police headquarters or the local magistrate's office, developing sources and tracking down leads. Over time, trust and understanding would grow between the reporter and the police officials with whom he or she came into contact. More recently, however, the frenzied reporting of such high-profile crimes as the O. J. Simpson and JonBenet Ramsey cases, coupled with a desire to satisfy the public's thirst for lurid and sordid details of a crime, has placed increased pressure on that type of relationship. For a police investigator, it is most important to strike a balance between the public's right to know and his or her need to maintain control of the information gathered during the investigation.

At the crime scene

The first contact between police and the media will undoubtedly take place at the scene of the crime. News organizations constantly monitor police radio communications and, in many cases, an investigating officer

who had rented the truck, from which artist's impressions were produced.

A massive operation was mounted by the FBI, employing over 1,000 agents armed with copies of the pictures, to question staff at gas stations, motels, and restaurants between Junction City and Oklahoma City, in the hope that someone would recognize one or both of the individuals. At the same time, the pictures were released to the media.

The campaign paid off. A hotel manager was positive that one of the pictures resembled a guest by the name of Timothy McVeigh, who had driven a Ryder truck. Then someone who had worked with McVeigh called the FBI to mention the similarity between him and the picture; he also indicated that McVeigh had been incensed by the Waco shootout.

When McVeigh's name was entered into the National Crime Information Center database, it revealed that he was already in Noble County Jail, having been arrested the day after the bombing for a traffic offense and carrying concealed guns.

McVeigh's driver's license gave a Michigan address, which also was the residence of Terry and James Nichols; Terry turned out to be the second bomber.

McVeigh was sentenced to death and executed in 2001; Nichols received life imprisonment.

will roll up to a scene that is already surrounded by television lights and cameras. In this situation, an investigator must assume control not only of the crime scene, but also the information that will be released. It is essential for an investigator or crime scene spokesperson to be aware that deadlines drive the media.

The early days of investigation

As a rule, during the preliminary stages of an investigation, the investigators are in possession of facts that absolutely cannot be shared with the media. While it is easy to restrict the information released by the police and other investigators, it is extremely difficult to control the media's access to individuals who were on the scene, and who may have witnessed the crime and possess information that may not even be known to investigators.

It is during the early, chaotic, and sometimes disorganized moments of an investigation that a slip of the tongue, the leaking of too much information, or the release of incorrect details can harm the investigation irreparably. Regardless of how large or small an investigative agency, having a designated media spokesperson is paramount. Many large police departments and agencies employ a full-time spokesperson, whose sole responsibility is to control the flow of information through press conferences. In small municipal police departments, this task often falls to a lead investigator or command-level official.

Public pressure for an arrest

In any homicide case, there is great community pressure to solve the case and make an arrest, but an investigator must resist the temptation to predict the outcome of an investigation. It is natural to want to ease tensions, and assure the public that police officials have the requisite skills and knowledge to solve a crime, but creating unreasonable expectations can have disastrous consequences. A rush to arrest with sketchy initial information may result in the apprehension of an innocent person, while making an arrest to satisfy community demands may shut down an investigation prematurely when more than one offender is involved.

In some cases, police make a public appeal for information through the media, but the decision to do so is weighed against the likelihood of undesirable consequences. Such an appeal

Colin Stagg is acquitted of the murder of Rachel Nickell in 1994, and is flooded with media attention outside the Old Bailey in London.

does reach a wider audience more quickly and more readily than a standard neighborhood canvass; it adds thousands of eyes and ears to the investigation, and may lead to the identity of a suspect or victim. On the other hand, a large amount of investigative time can be spent sifting through the deluge of information that often results, allowing the criminal to place time and distance between himself or herself and the crime. Consequently, important factors that are considered before making an appeal for information and assistance are the type of crime, and the size and character of the community.

Withholding details for the sake of the investigation

Regardless of the relationship that has been established between a police agency and the media, certain critical details of a case must be withheld to preserve the integrity of the investigation. An interview technique used to obtain a valid confession is to confront a suspect with information that only he or she would know about a crime. The veracity of such a confession could be compromised if the facts were widely known by the public. Restricting the amount of information released helps also to reduce the likelihood of copy-cat crimes being carried out.

08 The Trial

The culmination of all crime scene investigators' work is the trial. This is where all the evidence that police officers and investigators have gathered is used to prove beyond a reasonable doubt that the accused has committed the crime.

Usually within two weeks of the first appearance in court, or preliminary arraignment, the defendant will be back before the magistrate for the preliminary hearing. Unlike the one-party probable-cause review undertaken initially by the magistrate, this proceeding is adversarial, and both sides are represented by counsel. The burden of proof at the preliminary hearing remains one of probable cause, or *prima facie*, but the evidence is subjected to greater scrutiny and examination through the cross-examination of the prosecution witnesses by the defense counsel.

Although the rules of evidence for the conduct of preliminary hearings vary slightly between the United States, Canada, the United Kingdom, and Australia, in almost all cases, the prosecution is required to provide live testimonial evidence rather than evidence through affidavit. This is the first time the defendant will come face to face with the witnesses who will present evidence against him or her.

The prosecution is not required to present all the witnesses it plans to use in the eventual criminal trial. Typically, given the very low burden of proof required at the preliminary hearing, the prosecution will produce just

A U.S. courtroom where coroner's inquests are conducted.

enough evidence to establish that the crime was committed and that the person responsible was most likely the defendant. This is important for strategic reasons, because counsel for the defendant has the right to cross-examine the witnesses. A skilled defense attorney will use this opportunity to cast doubt on the credibility of witnesses and to develop impeachment testimony—statements that may be detrimental to the prosecution's case. At the preliminary hearing, the prosecuting attorney must strike a balance between presenting sufficient probable-cause evidence and protecting the credibility and integrity of the witnesses.

Once the prosecution has presented its case, normally that is the end of the preliminary hearing. Most jurisdictions allow the defense to present evidence at a preliminary hearing, but in many cases, that evidence must serve only as an effective rebuttal of the prosecution's *prima facie* evidence. Traditional defense strategies do not favor offering evidence for two reasons. First, it may expose possible defense arguments at trial; second, it may provide the prosecution with valuable impeachment tools at the time of trial—exposing any weaknesses of the defense witnesses far too early in the proceedings. If, for example, a witness testifies at a preliminary hearing that a car was red, and at trial that the car was blue, an attorney will confront the witness with the transcript of the inconsistent testimony.

At the conclusion of the evidence in the preliminary hearing, the magistrate has a number of options. He may determine that the prosecution has presented sufficient evidence to establish probable cause and will hand the matter over to the trial court. This is often the case because the evidentiary threshold is so low and the quality of evidence is usually high, especially in homicide cases.

Alternatively, the magistrate may find that probable cause does not exist for the crime charged, but does exist for a criminal offense of a lesser degree. In most jurisdictions, the rules of criminal procedure allow the prosecuting attorney to refile the original charges with a different magistrate if they disagree with the magistrate's decision.

Finally, if the magistrate finds that the prosecution's evidence fails to meet the burden of probable cause, the charges may be dismissed. In almost every case where a

Case file:

Timothy Spencer

In late 1987, four women were violently raped and strangled in Richmond, Virginia. There were strong similarities in the way all four crimes had been committed, and the police were convinced they were the work of one man.

All four victims were of similar appearance; all had been attacked while asleep; all had been bound and gagged using ropes and duct tape; all had been violently raped; and all had been strangled with ligatures. Notable aspects of the crimes were that the killer had been skilled at gaining access to the homes of his victims and clearly had not been fazed by the presence of other members of a victim's family in nearby rooms. The police thought it was likely that he was an experienced burglar. Hairs recovered from the crime scenes indicated that he was an African-American.

It appeared that the "South Side Rapist," as he had been dubbed by the media, had selected his victims carefully, and given the localized nature of the crimes, police sought a link between them to indicate why they had been chosen. Eventually, they realized that all four women had visited a particular shopping mall, and it was likely that this was the murderer's hunting ground.

Police mounted a surveillance operation and noticed that a known local cat burglar,

homicide is dismissed at magisterial level, the prosecuting attorney is certain to invoke the rules of criminal procedure and apply for an arrest warrant in a different jurisdiction.

The mountain of documents produced as evidence in the Edmond Safra murder trial in Monaco, 2002. The multibillionaire banker was asphyxiated in a locked bathroom of his Monte Carlo penthouse.

Pre-Trial Procedures

If a magistrate binds over a defendant for trial, a period between six months and a year may elapse before the defendant actually faces a judge and jury. Most states in America have speedy trial requirements in their rules of criminal procedure, and those defendants who are imprisoned are afforded the quickest opportunity to have their case go to trial. Similarly, in Canada, the United Kingdom, and Australia, anyone accused of a crime also has the right to stand trial within a reasonable time. This is the situation with most homicide defendants. Some jurisdictions provide that no more than six months may elapse from the time of arrest to the time of trial, and the

Timothy Spencer, spent time loitering in the mall. When he was pulled in for questioning on an unrelated burglary charge, police took blood, saliva, and hair samples for the recently introduced technique of DNA profiling. Police were astounded when the samples provided a genetic match with traces of semen found at the scenes of three of the murders. The odds of finding another match were considered to be 135,000,000 to one.

Spencer was sent to trial for the murder of the fourth victim, Susan Tucker. The DNA evidence formed a major part of the prosecution case. The defense tried to argue that any of his close relatives could have provided a match, but when the prosecution made it clear that they would introduce evidence tying Spencer to the other crimes, the defense retracted their argument. Spencer was convicted of Tucker's murder.

Subsequently, the combination of the crimes' "signature" and the DNA evidence convinced the jury that he had carried out the other killings. It also suggested that he had been responsible for two additional murders, which resulted in the freeing of someone who had been wrongfully convicted of one of them. Spencer was the first man in the United States to be executed largely on the basis of DNA fingerprinting.

failure of a prosecuting attorney or a governmental agency to observe this important requirement of due process may lead to the charges being dismissed, regardless of the strength of the case or the obvious guilt of the defendant.

While it may appear that the intense activity surrounding the case would probably die down after the arrest, arraignment, and preliminary hearing, the time between the preliminary hearing and the trial may be the most crucial. During this period, a wide variety of pretrial motions may be filed by the defense, in many cases raising objections to the validity of the proceedings that occurred between the arrest and the preliminary hearing. All of the previous actions of the case are now up for judicial review for the first time.

Pretrial motions commonly challenge the institution of the proceedings, allege deprivation of rights at the time of arrest, attack the sufficiency of the probable cause that led to the arrest, and request the suppression of evidence obtained by illegal means and in violation of a defendant's rights.

This is probably the most critical time in the criminal justice process for the investigating officer. The courts will review every aspect of the arrest and the subsequent handling of the suspect. Did the officer state sufficient probable cause for the arrest? Was the magistrate correct in issuing the warrant? Did the officer deny the suspect access to counsel? Did the officer properly ascertain whether or not the suspect was subject to custodial restraints when the first questions were posed? Did the officer use unlawful force or threaten to extract a confession? Was the confession voluntary, and was it reliable, given all the circumstances?

Of all possible pretrial motions, the most likely to succeed is the motion to suppress evidence. Quite often, if the defendant manages to suppress evidence that has been obtained by unlawful means, insufficient evidence will remain to allow the continued prosecution of the case.

The Criminal Trial

Many minor felonies and crimes of misdemeanor are disposed of without a trial. In these instances, the prosecution may propose an incentive to secure a guilty plea, frequently the dismissal of a more serious crime in exchange for a confession of guilt to a lesser offense. The prosecuting attorney may also offer a favorable sentencing recommendation to encourage a plea.

The courtroom scene during the trial of Washington, D.C. sniper suspect John Allen Muhammad. Defense witness Robert Holmes, Muhammad's friend, testifies during the penalty phase of the trial. Muhammad is seated at the left; attorney Paul Ebert addresses the audience.

In most homicide cases, however, a criminal trial is the rule rather than the exception. Little can be offered to secure a plea agreement when a defendant is faced with a substantial period of incarceration, life imprisonment, or, in the United States, the death penalty. At this stage of the justice process, the prosecution is faced with its highest evidentiary burden, that of proof beyond a reasonable doubt. And that must be established convincingly before a panel of jurors. Failure of the jury to reach a unanimous verdict will not necessarily result in the acquittal of a defendant, but rather the entry of a no verdict—more commonly known as a hung jury. In the case of a hung jury, the prosecutor must assess the strengths and weaknesses of the case, the availability and credibility of witnesses for a second trial, and the final jury tally before deciding to proceed with a retrial. If there are sufficient weaknesses and the witnesses have proved unreliable or lacking in credibility, and if only a few members of the jury favored conviction, the prosecutor may decide to withdraw the charges.

Other unique aspects to the criminal trial are the presumption of innocence, the right of the defendant not to take the stand, the prohibition of the jury from inferring anything by the defendant's failure to take the stand, and the exclusion of any evidence that was obtained by unlawful means.

A Day in the Life: Criminal Lawyer

Todd Edelman, Public Defender Service, District of Columbia, Washington, D.C.

When you are part of a team of lawyers that defends individuals charged with homicide, few days are alike. Each day requires that we focus on different issues, crises, and clients. Experienced lawyers at the Public Defender Service consider themselves fortunate to juggle a caseload of 20–25 pretrial serious felony cases at one time; other offices representing disadvantaged defendants expect attorneys to carry four times that number.

Most days begin in court. Even when attorneys are not in trial, they go to court to represent their clients in pretrial hearings. Some of these are simple administrative affairs that take minutes; others—including preliminary hearings, probation revocation hearings, and sentencings—involve extensive cross-examination and argument, and can last several hours. Often, an attorney will have several such hearings in a day.

After returning from court, attorneys working on homicide cases must spend their afternoons (and frequently evenings and weekends) preparing the groundwork for effective in-court advocacy. Charismatic and powerful courtroom performances certainly enhance an attorney's reputation; the work that allows those tactics to succeed, however, occurs months or even years before the trial, and usually out of public view. Defending a client facing a possible life sentence requires an attorney to pursue every available means of weakening the prosecution's case and securing the defendant's liberty. To build a successful defense, an attorney must research and write pretrial motions that may result in the exclusion of some of the prosecution's evidence; work with an investigator to locate witnesses who contradict aspects of the government's case, as well as information that undermines the credibility of the prosecution's witnesses; probe the strengths and weaknesses of the case against the defendant; and consult with expert witnesses to determine if a crime-scene photograph, bloodstain, or other trace evidence holds the key to securing a client's freedom. Perhaps most importantly, homicide attorneys must devote much of their non-court time to meeting with clients and clients' families.

For the majority of lawyers, trials are the most satisfying and challenging part of the job; the experience of trying (and winning) cases in front of juries makes even the worst aspects of our work seem worthwhile. Homicide trials often last several weeks and require single-minded focus not only in court, but practically around the clock.

Without question, defending a client charged with homicide involves a tremendous amount of work, stress, and expenditure of emotional energy. For most of us, the rewards we receive in court—working with brilliant and committed colleagues and helping our clients—more than compensate for any sacrifices we have to make.

Conduct of the Trial

The trial process begins with the opening statement, which is the first opportunity for the prosecution and the defense to tell the jury what the case is about. In almost all instances, the prosecution will present its opening statement at the beginning of the trial. The defense, however, may make its opening statement at the conclusion of the prosecution's opening, or it may wait until the the prosecution has presented its case. The defense attorney's decision will be based primarily on individual preference, although trial tactics also have a bearing.

Most lawyers, regardless of whether they are prosecutors or defense attorneys, will agree that the opening statement is the most important phase of a trial. Initial impressions have a profound effect on juries, and in many cases, the verdict is consistent with the impression made during the opening. Quite simply, first impressions are lasting impressions, and a lawyer must live with them throughout the course of a trial—whether it is a simple one-day affair or a weeks-long complex homicide case.

The most critical aspect of the opening statement is the clear, concise, and logical presentation of the theory of the case. Jurors are well aware that their function is to resolve a dispute. If the opening is confusing and long-winded, they will go through the trial without any understanding of where either side stands on the facts and the issues.

Unlike the United States, trials in the United Kingdom cannot be photographed or filmed. At left is an artist's sketch of the 1997 trial of teenager Nathan Brown, who was convicted of the murder of Carl Rickard with a machete outside his school.

Nearly 10 years after presiding over the high-profile case of former football star O. J. Simpson, Judge Lance Ito oversees a drive-by shooting murder case in May 2004.

Following the opening statements, the trial proceeds to the prosecution's "case in chief." This is the opportunity for the prosecutor to present all the facts of the prosecution's case in an easily understood manner, not only to explain the theory, but also to prove beyond a reasonable doubt that the defendant was responsible for the crime.

The prosecution's case

In some homicide cases, the prosecutor is fortunate to go to trial with a defendant's confession as the centerpiece of the case in chief. The homicide investigator will either testify to the facts and circumstances of the confession or will authenticate a video or audio recording of it. Confession testimony is a very difficult hurdle for a defense attorney to overcome, primarily because the validity and reliability of the confession are likely to have been established at the pretrial phase. Even the most experienced defense attorneys will have difficulty in discrediting confession testimony at trial, because the homicide detective offering the testimony is usually a veteran investigator with years of battle-hardened courtroom experience. With a confession as the crux of the case, the prosecutor needs only to produce enough additional evidence to establish the elements of the crime to achieve conviction.

In general, however, confession cases are exceptions to the rule, and in most instances the strength of the prosecution will lie in direct eyewitness testimony. This can be challenging for prosecutors, since eyewitnesses in homicide cases are usually not the most reliable or credible witnesses. In certain cases, the eyewitness may have an extensive criminal background, which will be subject to scrutiny and cross-examination by the defense attorney. Often, the eyewitness is a co-conspirator or an accomplice in the crime, whose motivation for offering testimony is easily discredited by the defense attorney.

Therefore, a successful prosecution is based on the way a prosecutor handles his witness. They should be the center of attention, and should be examined in such a way that nothing detracts from them. Witness credibility is determined by who they are, what they say, and how they say it. Even in the face of a less than credible background, the content of the witness's testimony and the demeanor in which it is presented may go a long way to securing a conviction.

At the conclusion of the prosecution's case, the trial moves into the defense phase. Initially, a defense attorney may ask for a judgment of acquittal, arguing that the prosecution, as a matter of law, has failed to establish the elements of a crime to the reasonable-doubt standard. Essentially, this is a request for the judge to take the case away from the jury and assume the role of fact finder in its place. A judgment of acquittal is rare in homicide cases, and succeeds only when the deviation from the burden of evidence is so apparent, or the cumulative testimony of witnesses is so deficient that no reasonable person could render a guilty verdict.

The defense's case

It is during this portion of the trial that the jury will hear for the first time positive defenses to the crime. Criminal defense practice dictates that a defendant does not take the stand in their own defense, nor is a defendant required to offer any testimony at trial. In addition, the jury is not permitted to draw any negative inferences from a defendant's failure or refusal to testify. Thus, the defense attorney uses a number of ploys to rebut the prosecution's case—bearing in mind all the while that the creation of at least a trace of reasonable doubt should result in an acquittal.

The defense attorney may suggest an alternate theory of the case, essentially attempting to disprove the conclusions of the homicide investigators and the forensic death investigators. Alternatively, the attorney may attempt to establish, through his or her own witnesses, the defense of self-defense, or possibly that the defendant has an alibi. The defense may also call witnesses to refute the eyewitness testimony of prosecution witnesses. In many cases, the defense will offer expert medical and forensic testimony that contradicts the cause-of-death evidence presented by the prosecution's forensic pathologist.

After each side's case in chief has been laid before the jury, the prosecution and defense make their closing arguments, and in most jurisdictions, the defense does this first. The closing arguments are the climax of the jury trial. They are the final opportunities for each side to communicate directly with the jury. Unlike the opening statements, however, closing arguments are exactly that—arguments. While the opening statement is a summation of the facts to be presented, the closing argument is completely the opposite. It encompasses all the logic of an opening, but does so at a highly emotional level. The most effective closing

Case file: George Russell

In the summer of 1990, a series of three murders occurred in Bellevue, Washington. All three victims were women, the first having been raped, and all three had been left naked in lewd poses. Moreover, the killer had incorporated various props into the poses. In the first case, the victim had been holding a large pine cone; in the second, a shotgun owned by the victim had been inserted into her vagina; in the third, the woman had been savagely slashed with a knife, a vibrator had been placed in the victim's mouth, and a copy of *The Joy of Sex* positioned in the crook of one arm. Various items had been used to cover parts of the bodies.

The first victim had been attacked on her way home late at night, had been strangled and subjected to several heavy blows and left in an alleyway. The other two had been asleep at home when the killer struck, beating both of them to death, in one instance when the woman's two daughters were sleeping in an adjacent room. These aspects of the crimes, the fact that only the first victim was raped, and variations in the pattern of wounds could have suggested two killers rather than one. However, the homicide investigators were convinced that all three killings were the work of one man.

A prime suspect in the case was George Russell, who already had a criminal record and

Counsel for the prosecution and defense convene at the judge's bench during the Paul Hill trial in Florida in 1994. Hill was convicted and executed for the shooting murder of an abortion doctor and his bodyguard.

had been arrested for brawling while impersonating a police officer. He had been found in possession of a gun that had been stolen from a home close to the scene of the third crime. He was released, but subsequently caught when police were called to investigate a prowler seen in Bellevue at night. He was arrested again, but this time the police took DNA samples; they matched traces found on the first victim.

An FBI profiler was called in to examine the evidence of the three crimes, and he found a significant number of similarities between them. These included the way in which the killer had carried out rapid, overwhelming attacks on the victims; the fact that he had left his victims where they would be found quickly; the degrading poses in which he had placed them; and the use of props to emphasize his fantasy. This signature evidence was sufficient to convince the court that Russell should be tried for all three murders.

Robert Keppel, an expert on serial killers, was summoned by the prosecution to testify at Russell's trial. Keppel described the aspects of a signature killing and successfully linked them to the three murders under review. Their standard pattern clearly implicated George Russell in all three cases.

Russell was found guilty on all three counts and received a life sentence.

argument will persuade a jury to reach a conclusion and feel good about it afterward. In the prosecution's case, the ultimate conclusion is for the jury to render a guilty verdict after considering all the evidence that proves the crime beyond a reasonable doubt.

Proof Beyond a Reasonable Doubt

The highest standard of all evidentiary burdens is reserved for criminal trials—the burden of proof beyond a reasonable doubt. For the prosecution to secure a conviction in a criminal trial, the burden of persuasion must exceed all other burdens of proof at earlier stages of the proceedings. It is up to the prosecutor to meet the burden of proving every element of the crime that is charged, and to convince the trier of fact, whether judge or jury, that each element exists beyond a reasonable doubt. The prosecutor must also establish beyond reasonable doubt the defendant's participation in the crime or his or her responsibility for it.

The reasonable-doubt standard begins with the premise of all criminal prosecutions—that the accused is presumed to be innocent, and the fact that he has been charged with a crime should not be taken as an indication of his guilt.

Most judicial systems have long viewed the application of the reasonable-doubt standard as an indispensable instrument for reducing the risk of convictions that rest on factual error. There are many reasons why the reduction of the margin of error is critical. First and foremost, the defendant's liberty is at stake. A conviction could not only ensure the loss of liberty—possibly for life in the case of a homicide, but also severely stigmatize a person following release from imprisonment. Second, the faith the public holds in the criminal justice system would be seriously eroded if there was any suggestion that innocent people were being convicted routinely. Recent history is littered with overturned convictions following advances in scientific methods, yet obviously the reasonable-doubt standard was met originally for those convictions to have been achieved. Finally, every individual should have confidence that they cannot be judged guilty of a criminal offense without some fact finder being convinced of guilt with the utmost certainty.

LEFT A map of the Washington, D.C. sniper shooting sites was used during testimony against John Allen Muhammad.

BELOW A police forensics specialist holds the rifle scope found in Muhammad's car.

Forensic Scientific Expert Witnesses

Forensic pathologists, testifying to the cause and manner of death in homicides, and forensic psychiatrists, testifying in criminal cases where the insanity defense has been raised, have been leaders in the development of forensic medicine in the United States and Canada, as well as in many other nations of the world, particularly in Western Europe. The forensic medicine specialist is now one of many forensic scientists—including the toxicologist, the immunologist, the odontologist, the criminalist, and the anthropologist—who provide valuable services in courts of law.

Increasing in importance also is the clinical medical specialist, who is frequently called upon to give medical evidence in court. This specialty includes not only the orthopedic surgeon, who is most often called to testify in personal injury matters, but also other medical specialists who may be asked to testify less frequently, such as emergency room physicians (trauma cases) and pediatricians (child abuse cases).

Scientific and expert evidence

While a forensic pathologist may be able to ascertain certain information from his examination of the human body, a trained criminalist may be able to reach additional conclusions by a thorough analysis of the victim's clothing. The degree of confidence that may be placed in these two experts is related directly to their experience, their training, their credentials, and whether they have performed all necessary tests. For example, in determining whether a particular weapon was used in a crime, an attorney will want to know if the criminalist fired the exact weapon into cotton wadding and onto paper, and if any tests were performed on animal carcasses or other materials that would simulate the human body. This is because a specific weapon will not necessarily discharge in the same fashion as others of its type.

Case file: Jean Harris

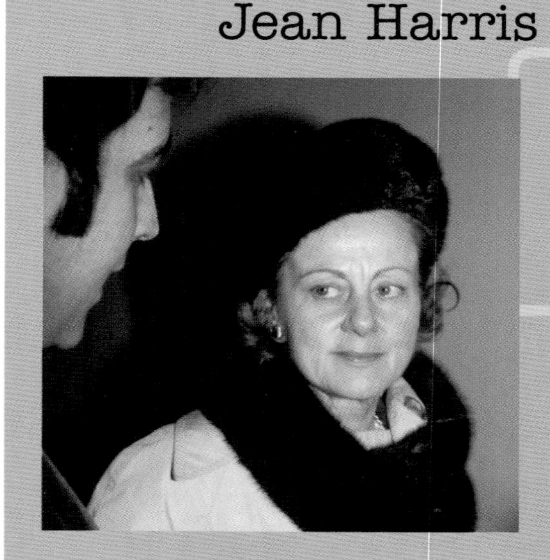

Jean Harris and Dr. Herman Tarnower had been lovers for 14 years, but by early 1980 he was beginning to tire of her. Tarnower, 69 and a renowned cardiologist, had developed the Scarsdale diet; Harris was the 57-year-old headmistress of a prestigious girls preparatory school who suffered from depression and was racked with self-doubt about her relationship with Tarnower because of her age. She knew, for example, that Tarnower was in a long-term relationship with a younger woman, Lynne Tryforos, a nurse-receptionist at the Scarsdale Clinic.

Matters came to a head when Tarnower invited both women to a banquet in his honor.

The credibility of the expert witness

Attorneys must distinguish between forensic scientific experts who will provide the most accurate information and those who may be influenced by bias. For example, if a forensic pathologist is a novice or becomes too closely allied to the police, and begins to think that he or she is an adjunct of the prosecution, that forensic pathologist may be dangerous on the witness stand. In this instance, a defense attorney must tread warily. Ideally, the defense would want some advance idea of what the expert witness's answers were going to be, even if they were prepared to bring in their own expert and merely intended to challenge the witness's credibility.

Medical vs. Legal Concepts of Proof

The scientist and the physician require a higher level of evidence to prove cause than the court. Expert witnesses from these disciplines carry these more stringent demands into the courtroom, where they continue to think in terms of scientific proof, rather than the probabilities the law would have them consider.

Harris felt slighted and wrote him a long, caustic letter, in which she called Tryforos a slut, accused her of damaging clothes Harris had left at Tarnower's home, and called her a "vicious, adulterous psychotic."

With her world collapsing around her, Harris decided to commit suicide, but wanted to see Tarnower one last time. On March 10, she drove to his home in Westchester County, N.Y., taking with her a .32-caliber revolver. Arriving at 10:00 P.M., she found him asleep in bed. She woke him and they argued heatedly, during which he was shot four times, fatally.

At her trial, Harris's defense claimed that she had intended to shoot herself and that Tarnower's death had been a tragic accident. They referred to her bouts of depression and called character witnesses on her behalf.

Harris, however, appeared to show no remorse for what had occurred, and her description of events at Tarnower's home lacked conviction. The vital piece of evidence, though, was the letter. This was used by the prosecution to debunk the defense claim that Harris was a respectable woman who could not have deliberately murdered her lover.

Harris was convicted of second-degree murder and sentenced to 15 years to life. While in prison she published three books and gave parenting classes to other inmates.

Former National Basketball Association player Jayson Williams is acquitted of the serious charge of aggravated manslaughter in the shooting death of Costas "Gus" Christofi but is found guilty on four counts, including evidence and witness tampering.

Medical proof in the laboratory is complex and tends to implicate multiple factors or mechanisms. It is quite rigid and strict, and demands indisputable substantiation to prove a causal relation. In contrast, legal proof is much more practical. It focuses on probable or legally effective causation. Its goal is the determination of legal responsibility.

While attorneys, judges, and juries may be content to make judgments that, more often than not, serve the cause of justice, the scientist requires more than reasonable certainty if his work is to have scientific validity.

Even when the cause of a disease is unknown or obscure, the law may still be content to accept testimony on causation as sufficient to support a verdict in favor of an injured person, although this may not seem scientifically credible. This is acceptable to the legal system because its purpose is to settle disputes. When the cause is unknown, the law must still decide whether the plaintiff or defendant should suffer the loss. If the plaintiff wins, this does not mean that the law considers the cause to have been established as a scientific truth. Thus, the trauma a defendant inflicted upon a plaintiff may be proved to have been the legal cause of the plaintiff's cancer; it does not mean that trauma generally is a scientific cause of cancer.

The physician searches for the basic or underlying cause; the attorney attempts to show that a particular event hastened, precipitated, or aggravated the condition. The physician looks at every aspect of his or her patient; the attorney looks only to the injured condition that is the subject of the legal action. The physician demands objective data; the attorney may use any admissible evidence available and will settle for reasonable medical certainty, reasonable probability or, in some instances, even a mere possibility. These and other differences frequently lead to a significant breakdown of communication between the two professions.

There can be ambiguity in the words themselves. The medical expert who says "possible" may mean that a significant likelihood exists. A physician is less likely to use the word "probable," since that may mean near certainty to them, rather than a 51 percent likelihood. If a physician feels there is a 60 percent likelihood that the plaintiff's injury will lead to a particular complication, they might call the complication "possible" if asked; if the court knew that the physician's "possible" meant a 60 percent chance, it might characterize this as "reasonable medical certainty," rather than a mere "possibility."

The jury's function is to resolve conflicts, and where an expert's testimony is undisputed and certain regarding causation, there is no conflict to resolve. Thus, the jury should logically be bound by the expert's opinion. If, however, the testimony falls short of certainty, the jury will be required to weigh its credibility.

San Diego County Medical Examiner Brian Blackbourne, MD, gives expert testimony on the witness stand during the preliminary hearing for David Westerfield, who was charged with the kidnap and murder of Danielle van Dam.

From a juror's perspective

On the one hand, jurors sometimes feel they are being asked to make a decision they are not qualified to make, especially in complex cases. They are grateful to experts who explain case issues simply, clearly, and in a way that helps them make a reasonable determination. On the other hand, jurors are disturbed by the concept that an attorney can hire someone to back up almost any position. This concern is reinforced when both sides bring out highly qualified experts who are presumed to know everything there is to know about the subject in question, but who contradict each other on every point. The notion of the hired gun runs counter to jurors' sense of justice, and they are likely to question an expert's motive for testifying.

What jurors want from experts is an explanation. If they feel they are getting an argument instead, they tend to counter this in their own minds, and resist the expert's efforts to be persuasive. For persuasion to work, it must be self-persuasion. In other words, it is what listeners tell themselves that is most compelling. If they feel an expert should not be advocating in the first place, jurors tend to tell themselves why the expert is wrong and should not be acting like a lawyer. As a result, the jurors will not accept what the expert is saying. Experts who are perceived as not being invested in the outcome of the case and who are not overly aggressive, arrogant, or defensive on the stand are considered more credible. Emotional investment with uncontrolled zeal or passion can prove damaging.

Case file:
O. J. Simpson

On June 12, 1994, officers from the Los Angeles Police Department were called to the home of Nicole Brown Simpson, former wife of renowned American football player O. J. Simpson. They found the bodies of Mrs. Simpson and her friend, Ron Goldman, both stabbed to death.

Investigators discovered a wealth of forensic evidence at the scene: both bodies covered in blood, blood on a gate, bloody shoe prints, and a bloodstained left-hand glove.

Suspicion fell on O. J. Simpson, who tried to flee by car. Police vehicles gave chase, while the whole episode was being filmed by TV news helicopters. Eventually, Simpson gave himself up.

Medicine is a combination of art and biological science. Consequently, reasonable persons may differ in their opinions within the realm of medicine. The extent of disagreement varies with the nature of the problem and the specialty involved. For example, there should not be much argument about the conclusions of an experienced forensic pathologist regarding the trajectory of a gunshot wound, or the location and size of a subdural hematoma (blood clot in the skull). On the other hand, a great deal of subjectivity could be involved in the opinion of a psychiatrist about whether or not a defendant is competent.

Because reasonable minds may differ in such instances, there is nothing wrong with an intelligent, probing cross-examination designed to point out weaknesses and inaccuracies in the conclusions and opinions of the other side's expert witnesses. And if opposing counsel genuinely believes the expert is deliberately lying or distorting known facts and fundamental scientific truths, it is understandable and acceptable that cross-examination may have to be tough, brutal and even vicious.

When forensic tests were carried out on the evidence, the case against Simpson appeared overwhelming. The blood on the gate produced so close a match to his DNA that only one person in 57 billion was likely to have had the same genetic profile. The shoe prints were Simpson's size and came from an unusual designer shoe he had been seen wearing. When arrested, he had a cut on his left hand, while a right-hand glove that matched the glove from the scene was discovered outside his home. Traces of the victims' blood were in his car and home.

Despite the amount of forensic evidence, Simpson was acquitted. The defense argued that officers handling the case were racist and cast doubt on the way the forensic tests had been conducted. Witnesses swore they had seen Simpson on the day of the murders, at a time that would have prevented him from carrying them out. There was a suggestion that a second killer could have been present. Sufficient doubt was sown in the minds of the jurors that they found Simpson not guilty.

That was not the end of the story, though. Relatives of the victims brought a civil prosecution against Simpson that found him responsible for both deaths. Damages of $33.5 million were awarded against him, but Simpson was broke, and the money is yet to be paid.

About the Authors and Contributors

Authors

Joseph T. Dominick, RN, LFD

Chapter 1
Chief deputy coroner for the ACCO, Joseph Dominick has served as a death investigator since 1990. In addition to his duties at the ACCO, he is a practicing emergency room nurse, and a licenced funeral director and embalmer. He is an adjunct professor of forensic nursing at Duquesne University and has lectured extensively on death investigation and forensic nursing.

Steven A. Koehler, MPH, PhD

Chapters 1–6
A forensic epidemiologist, Steven A. Koehler received his masters and doctoral degrees in forensic epidemiology from the Graduate School of Public Health at the University of Pittsburgh. He is employed by the Allegheny County Coroner's Office (ACCO), where he carries out statistical analysis for research projects and for use in publications. He has contributed to articles addressing such topics as cardiac concussions, deaths during police procedure, medical misadventures, epilepsy, and Kawasaki disease. He also holds seats on the Suicide Death Review committee, the Firearm Death Review board and the Internal Monitoring Committee for the Center for Injury Research and Control. He is an adjunct associate professor of epidemiology at the University of Pittsburgh, and teaches at local colleges, hospitals, and the University of Pittsburgh Graduate School of Public Health.

Shaun Ladham, MD

Chapters 3 and 5
Forensic pathologist Shaun Ladham attended medical school at Dalhousie University in Halifax, Nova Scotia, and received residency training at McGill University in Montreal, Quebec. In addition to his duties at the ACCO, Dr. Ladham sits on the Allegheny County Child Death Review Board, the Suicide Death Review Board, and HIV advisory panel. He participates in surgical trauma rounds at Allegheny General Hospital and Children's Hospital of Pittsburgh. He is a member of the National Association of Medical Examiners and Canadian Association of Pathologists.

Thomas Meyers, BS, MS

Chapter 4
Tom Meyers holds a Bachelor of Science degree in Chemistry and a Master of Science degree in Forensic Chemistry from the University of Pittsburgh. In 1977, he joined the then Allegheny County Crime Laboratory as a criminalist, and was assigned to the serology unit. He completed DNA training through the FBI Academy in 1989, assumed the position of DNA technical leader in 1995, and became DNA/serology supervisor in 1997. He completed DNA auditor training in 1998 and is currently a

member of the Mid-Atlantic DNA Auditing Group. Tom Meyers lectures at the University of Pittsburgh School of Medicine on molecular genetics and its application to forensic science.

Timothy G. Uhrich, JD

Chapter 7

Attorney for the ACCO, Timothy Uhrich also serves as a deputy coroner. In addition, he maintains a private practice concentrating on family law, estate planning and administration, and general law. Uhrich is a 1982 cum laude graduate of the University of Pittsburgh. He received a juris doctor from the Duquesne University School of Law in 1988. He has extensive experience in government and public service, and has acted as an assistant city attorney for the City of Pittsburgh. He also served in the United States Army reserves, and was commissioned in 1972 as a Second Lieutenant in the Military Police Corps.

Cyril H. Wecht, MD, JD

Introduction and chapter 8

Coroner of Allegheny County, Cyril H. Wecht is also one of the United States' leading forensic pathologists. He is certified by the American Board of Pathology in anatomic, clinical, and forensic pathology, and is a Fellow of the College of American Pathologists and the American Society of Clinical Pathologists.

Dr. Wecht is a clinical professor at the University of Pittsburgh Schools of Medicine, Dental Medicine, and Graduate School of Public Health, and holds positions as an adjunct professor at Duquesne University Schools of Law, Pharmacy, and Health Sciences. He has served as president of the American College of Legal Medicine, the American Academy of Forensic Sciences, and as chairman of the Board of Trustees of the American Board of Legal Medicine and the American College of Legal Medicine Foundation.

Dr. Wecht has organized and conducted postgraduate medicolegal seminars in more than 50 countries around the world in his capacity as director of the Pittsburgh Institute of Legal Medicine. He has performed approximately 14,000 autopsies, and has supervised, reviewed, or has been consulted on approximately 30,000 additional postmortem examinations.

An expert in forensic medicine, Dr. Wecht has frequently appeared on nationally syndicated television programs to discuss medicolegal and forensic scientific issues, including medical malpractice, drug abuse, the assassinations of President John F. Kennedy and Senator Robert F. Kennedy, the death of Elvis Presley, the O. J. Simpson case, and the JonBenet Ramsey case. His expertise has also been employed in a number of other high-profile cases. The author of more than 400 professional publications, Cyril H. Wecht is an editorial board member of more than 20 national and international medicolegal and forensic scientific publications. He is the author of *Cause of Death*, *Grave Secrets*, and *Who Killed JonBenet Ramsey?*

Michael Welner, MD

Chapter 6

Michael Welner is known for many innovations that have influenced the practice of forensic psychiatry. As chairman of the Forensic Panel, he founded the first peer-reviewed forensic expert consultation practice in the country. Recently, he developed the Depravity Scale, a device currently being validated for the psychological determination of evil. A clinical associate professor of psychiatry at New York University School of

Medicine and an adjunct professor of law at Duquesne University, Dr. Welner is supervisor to trainees in psychiatric diagnosis, forensic psychiatry, psychopharmacology, and treatment decision making. Honored by the American Psychiatric Association in 1997 for excellence in medical education, he has acted as a consultant for both prosecution and defense counsel in a number of high-profile cases.

"A Day in the Life" Contributors

Paul Anderson

Paul Anderson is the longest-serving police reporter at the Melbourne newspaper the *Herald Sun*. He was part of a team that won a prestigious Australian national Walkley Award, and has been highly commended in Victoria's Quill Awards. He is the author of two true-crime books.

Nicola Aranyi

Nicola Aranyi is a principal analyst for Lincolnshire Police in the United Kingdom. Previously she was with the Hertfordshire Constabulary as an analyst from 1997 to 2002.

Todd Edelman

Todd Edelman is currently a visiting professor in Georgetown University's Criminal Justice Clinic, on leave from the Public Defender Service for the District of Columbia, where he has been an attorney since 1997 and chief of the Serious Felony Section since 2002. Edelman has represented dozens of defendants charged with homicide.

Dr. Frederick W. Fochtman

Dr. Fochtman is chief toxicologist at the ACCO. He is associate professor of pharmacology and chairman of the Department of Pharmacology-Toxicology in the School of Pharmacy at Duquesne University in Pittsburgh.

Robert B. Kennedy

Robert Kennedy is a world leader in barefoot morphology comparisons. A Sargeant in the Royal Canadian Mounted Police, he has taught and testified in cases around the world, as well as published widely on tire and foot impressions in forensic science.

Dr. Katherine Ramsland

Dr. Ramsland teaches forensic psychology at DeSales University in Pennsylvania, and is the author of 20 books, including *The Forensic Science of CSI* and *The Criminal Mind: A Writer's Guide to Forensic Psychology*.

Dr. Kim Rossmo

Detective inspector Kim Rossmo's doctoral dissertation in 1995 at Simon Frasier University's School of Criminology was groundbreaking in establishing the field of geographical profiling and has been used extensively in Canada, the United States, and the United Kingdom. The first geographical profile was produced by Rossmo's Vancouver police department in 1990.

Dr. Michael N. Sobel

For over 30 years, Dr. Michael Sobel has been the chief forensic odontologist for the ACCO. Dr. Sobel has appeared on television, and has written and presented many chapters and papers in his forensic fields. His trial experience is extensive, and he regularly acts as a forensic odontology consultant for both prosecution and defense counsel throughout the United States.

John J. Van Tassel

Corporal John Van Tassel has been with the North Vancouver Forensic Identification Section since 1989. An expert knot analyst, Van Tassel has worked on the high-profile JonBenet Ramsey case, among others.

Dean A. Wideman

Dean Wideman is a consulting forensic scientist and criminal profiler at NucleoGenix, a forensic services company based in San Antonio, Texas. He is the former executive director and chief forensic scientist and criminal profiler of Crime Analytica, Inc., a nonprofit organization that provides support services to victims of crime and their families.

Acknowledgments

The authors gratefully acknowledge the assistance of Marty Coyne, David A. Crown, and Neal H. Haskel in the preparation of the text for this book; also of Marty Coyne and Lisa Leon in providing many of the photographs.

The publishers would like to thank the following for generously contributing to the "Day in the Life" pieces: Paul Anderson, Nicola Aranyi, Marty Coyne, Joseph T. Dominick, Todd Edelman, Dr. Frederick W. Fochtman, Robert B. Kennedy, Dr. Steven A. Koehler, Dr. Shaun Ladham, Dr. Kim Rossmo, Dr. Michael N. Sobel, John J. Van Tassel, Dean A. Wideman.

Thanks also to Thomas Keenes for design assistance, and Roger Forsdyke and Martin Edwards for expert advice.

Picture Credits

Allegheny County Coroner's Office: 17, 18, 23, 26, 30, 31, 32, 35, 41, 43, 45, 46, 49, 59 (top left, top right, and bottom), 63, 64, 65, 66, 67, 68 (top and bottom), 69, 70 (top and bottom), 72, 81, 84, 85 (top and bottom), 89, 92, 93, 100, 102, 103, 106 (left, center, and right), 113, 114, 115, 116, 120, 124 (top), 153, 167

Corbis: 10, 13, 14, 20, 37, 44, 98, 122, 128 (top and bottom), 132, 138, 142, 149, 180, 183, 184

Getty Images: cover (all 3 images), 9, 12, 27, 54, 137, 140, 154, 156, 174, 179 (left and right), 182

Rex Features: 28, 39, 52, 62, 74, 79, 86, 127, 134, 144, 161, 162, 165, 169, 171, 173, 177

Science Photo Library: 42, 51, 57, 73, 77, 90, 97, 107, 111, 124 (bottom)

Index

Page numbers in *italic* type refer to illustrations.

A

B

C

D

E